Advances in Polymer Science

Fortschritte der Hochpolymeren-Forschung

Volume 16

Edited by

H.-J. Cantow, Freiburg i. Br. · G. Dall'Asta, Milano · J. D. Ferry, Madison
H. Fujita, Osaka · M. Gordon, Colchester · W. Kern, Mainz · G. Natta, Milano
S. Okamura, Kyoto · C. G. Overberger, Ann Arbor · W. Prins, Syracuse
G. V. Schulz, Mainz · W. P. Slichter, Murray Hill · A. J. Staverman, Leiden
J. K. Stille, Iowa City

With 36 Figures

Springer-Verlag Berlin Heidelberg GmbH 1974

Editors

ISBN 978-3-662-15939-2   ISBN 978-3-540-37860-0 (eBook)
DOI 10.1007/978-3-540-37860-0

The Entanglement Concept in Polymer Rheology

WILLIAM W. GRAESSLEY

Chemical Engineering and Materials Science Departments, Northwestern University, Evanston, Illinois 60201, U.S.A.

Table of Contents

1. Introduction

The concept of chain entanglement first arose more than 40 years ago from attempts to explain the mechanical properties of amorphous polymers above the glass transition temperature. In 1932 Busse (*1*) noted that if unvulcanized rubber is subjected to a large deformation, held for a short time and released, it recovers its original shape almost completely, while if held for a long time it flows and recovers only partially when released. He distinguished between the weak van der Waals forces between molecules (flexible fibrous units in his terminology) over most of their length which offer almost no resistance to motion, and a few widely-separated strong interactions which serve, for short times at least, to bind the structure into a three-dimensional network. He attributed these strong interactions to a physical interlocking of the molecules which can slip to new equilibrium positions if given time, and he distinguished such temporary links from the permanent chemical linkages provided by vulcanization.

In 1940 Treloar (*2*) pointed out that such physical coupling or entanglement might indeed be expected, given the long flexible nature of individual molecules and a random, interpenetrating arrangement in the solid. Regions where molecules were looped through one another might offer high resistance to deformation for a time, but the loops would eventually slip or be removed and reformed by random thermal motion. He commented further that most of the observations on unvulcanized rubber—large, recoverable deformations for short times, stress relaxation and viscous flow for long times—could be satisfactorily explained if one regarded entanglements as isolated regions of high viscosity, interconnected by freely extensible molecular segments.

In 1944, Flory (*3*) noted that the moduli of cross-linked butyl rubbers generally differ somewhat from values calculated from the crosslink density according to the kinetic theory of rubber elasticity. In many cases, the modulus also depends on the primary (uncross-linked) molecular weight distribution of the polymer. He attributed both observations to three kinds of network defects: chain ends, loops, and chain entanglements. The latter are latent in the system prior to cross-linking and become permanent features of the network when cross-links are added.

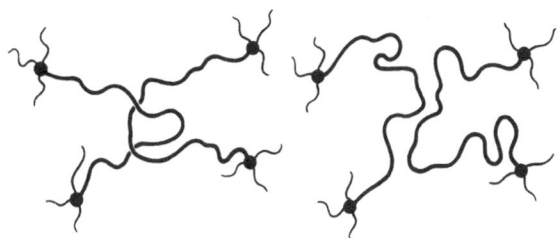

It was argued that such structures would behave as additional network constraints, and thereby increase the equilibrium modulus of the network.

In 1946 Green and Tobolsky (4) considered the properties of a rubberlike network in which the junctions are being continually broken and reformed. They showed that under certain circumstances the stress could be expressed as a function of the history of the strain, and that such typical viscoelastic properties as viscous flow and stress relaxation were natural consequences. Buchdahl (5) in 1948 suggested that the shear rate dependence of viscosity in polymer melts could be accounted for by a net reduction of entanglement coupling in steady flow. Nielsen and Buchdahl (6) in 1950 estimated M_e, the molecular weight between coupling points, using creep and stress relaxation data and the modulus equation from the theory of rubber elasticity. For polystyrene melts they obtained $M_e = 10\,000 - 40\,000$.

In 1952 Bueche (7) proposed the first model for the extra frictional drag conferred by entangling interactions and used it to calculate the molecular weight dependence of melt viscosity for linear chains. The result [as revised in 1956 (8)] was $\eta_0 \propto M^{3.5}$ for highly entangled chains, in good agreement with the empirical relation $\eta_0 \propto M^{3.4}$ (9, 10), found at high molecular weights.

In 1955 Ferry, Landel, and Williams (11) applied the Rouse-Bueche molecular model to concentrated systems, and suggested that the largest relaxation times (those associated with the motion of portions of the chain longer than the entanglement spacing) were simply shifted to longer times by entanglement drag, the short relaxations being essentially unaffected. According to this view the plateau region in stress relaxation is created by a simple splitting of the relaxation time distribution. The plateau modulus for high molecular weight polyisobutylene gave $M_e = 6700$. It was noted that this value was smaller but nevertheless of the same order of magnitude as the molecular weight at the break in the η_0 vs M curve ($M_c = 17\,000$ for polyisobutylene).

In 1956, Lodge (12) and Yamamoto (13) showed that a wide range of optical and mechanical properties of deforming polymers could be described quantitatively through the concept of a transient network of flexible random coil elements, the junction points being specifically identified as chain entanglements. A mechanical constitutive equation for such networks was developed, and a theoretical basis was laid for flow birefringence as a means of evaluating mechanical properties in concentrated polymer liquids (12, 13a). In 1958, Bagley and West (13b) pointed out the close connection between prominent shear rate effects in the viscosity and the presence of entanglements.

During this same period, the equilibrium stress-strain properties of well characterized cross-linked networks were being studied intensively. More complex responses than the neo-Hookean behavior predicted by kinetic theory were observed. Among other possibilities it was speculated that, in some unspecified way, chain entanglements might be a contributing factor.

Two major developments since the mid-1950's have greatly influenced the entire field of rheology. First, continuum theories have shown that, subject to very broad and physically plausible assumptions about mechanical response, many seemingly independent viscoelastic properties of a material are in fact related. These relationships have been confirmed experimentally and should hold regardless of the specific molecular mechanisms for relaxation and flow. Second, a considerable body of viscoelastic data has been accumulated on well characterized polymers in concentrated solutions and in bulk. The studies show that properties which depend principally on the longest relaxation times of the system vary systematically with concentration, molecular weight and molecular weight distribution. Broadly speaking, the manner of these variations is independent of the chemical structure of the repeating unit.

Molecular theories, based in large part on ideas about chain entanglements, have been constructed to explain certain of these observations. The theories must still be regarded as tentative and incomplete. They are based, first of all, on reasonable but still incompletely accepted ideas about chain organization in concentrated solutions and melts. Secondly, they deal with the response of individual chains or pairs of chains in a smoothed medium, rather than with an entire interacting ensemble. Finally, they circumvent the deep mathematical difficulties of the central problem, interaction between mutually "uncrossable" sequences of chain elements, by approximations which are not easy to evaluate. The purpose of this review is to summarize the present status of entanglement theories and the data upon which they are based.

Chain entanglement is probably best regarded as a special type of intermolecular interaction, affecting mainly the large-scale motions of the chains, and through them, the long time end of the viscoelastic relaxation time spectrum. Evidence for the existence of entangling interactions comes entirely from mechanical properties, specifically, the flow properties of solutions and melts and the equilibrium mechanical properties of networks. The effects appear most prominently at high polymer concentrations and high molecular weights in polymer liquids and large primary chain lengths and low cross-linking densities in the case of networks. Molecular weight distribution and long chain branching influence all flow properties, profoundly in some cases, and their effects are especially important in the entanglement regime. Data are accumulating on these influences, and a critical review of them would undoubtedly be useful. However, to keep the current review to a manageable length, and to focus attention on the entanglement phenomenon itself, we will deal here mainly with the behavior of linear polymers with narrow molecular weight distribution, making only occasional reference to polydispersity and branching effects. For similar reasons we will omit discussion of elongational flows and behavior observed in more complex flows, such as die swell.

Quantitative evidence regarding chain entanglements comes from three principal sources, each solidly based in continuum mechanics: linear viscoelastic properties, the non-linear properties associated with steady shearing flows, and the equilibrium moduli of crosslinked networks. Data on the effects of molecular structure are most extensive in the case of linear viscoelasticity. The phenomena attributed to chain entanglement are very prominent here, and the linear viscoelastic properties lend themselves most readily to molecular modeling since the configuration of the system is displaced for equilibrium only slightly by the measurement.

In steady shear flows, only the shear-rate dependence of viscosity is well documented in terms of molecular structure effects. Molecular theories are more difficult because the configuration of the system is appreciably displaced from equilibrium. Indeed the simplest and most attractive molecular models of linear visoeleastic behavior predict no shear rate dependence whatsoever for the viscosity. Entanglement concepts enter here in a more indirect and qualitative way: to explain the existence of a prominent dependence of viscosity on shear rate at high concentrations and molecular weights, and to account for and predict the forms of the other non-linear properties. *One must, of course, be careful not to attribute more than necessary specifically to chain entanglement.* The viscosity of polymer systems depends on shear rate even when intermolecular interactions are absent, and very simple molecular models predict elasticity in polymer liquids without additional assumptions about entanglement.

Modulus data on crosslinked systems would seem to offer the most direct method for studying entanglement effects. Certainly, from the standpoint of molecular modeling, the advantages of equilibrium properties are clear. However, the structural characterization of networks has proven to be very difficult, and without such characterization it is almost impossible to separate entanglement contributions from those of the chemical crosslinks alone. Recent work suggests, however, that these problems are not insurmountable, and some quantitative results are beginning to appear.

The sections on experimental results from linear viscoelasticity (Section 5), entanglement theories in linear viscoelasticity (Section 6), entanglements in networks (Section 7) and non-linear viscoelasticity (Section 8) comprise the major portion of the review. Section 2 concerns the current evidence on configuration and arrangement of chains in concentrated polymer liquids. Section 3 is a brief summary of phenomenological equations in linear viscoelasticity and steady shearing flow. Section 4 is a summary of molecular models, principally the spring-bead models, which provide the framework for discussing entanglement effects. However, in all cases involving experimental data we have tried to carry the correlations as far as possible before introducing specific molecular interpretations.

This article builds upon an earlier review on the same subject by Porter and Johnson in 1966 (*14*), and on the recent treatise on viscoelasticity in polymers by Ferry (*15*). We have generally tried to maintain the same nomenclature as the latter. Recent reviews on the relation between the zero-shear viscosity and molecular structure (*16*), crosslinked networks (*17*), and flow birefringence (*18*) in this same journal cover portions of the subject. We have tried to minimise redundancy with these works while at the same time making the review reasonably self-contained.

2. Chain Configuration in Amorphous Polymer Systems

2.1. General Considerations

Random crosslinking reactions lead quite efficiently to network formation, requiring that the individual chains have relatively open configurations which provide plentiful intermolecular contacting. On the other hand, crosslinked networks can undergo large and reversible deformations, a property which demands a rest state of relatively unextended configurations for the network strands. The random coil configuration, such as that observed in dilute solutions, fulfills both requirements. The volume fraction of segments from any individual chain is small throughout its own pervaded volume. Since the space in concentrated systems is densely filled with polymer segments, random coils must overlap and interpenetrate extensively. Intermolecular contacts are therefore plentiful for all repeating units in each chain. At the same time, the end-to-end distance is small compared to contour length in random coils, so considerable extension is possible without the breaking of chemical bonds.

Despite its plausibility and the apparent absence of driving forces to produce other large scale arrangements, the random coil configuration has been questioned on various grounds. It seems worthwhile to review the evidence at this point, since most molecular theories for amorphous polymers are based on the random coil picture.

2.2. Large Scale Configuration in Concentrated Systems

2.2.1. Theoretical Aspects

The random coil model is firmly established for flexible, linear molecules of high molecular weight in dilute solution (19, 20). Characteristic dimensions such as the mean end-to-end distance and radius of gyration are controlled by both local chain structure (bond angle and steric restrictions) and long-range interactions. The latter arise from the mutually excluded volume of segments widely separated along the chain contour. The magnitude of the perturbation depends also on the thermodynamic interaction between polymer segments and solvent. The net perturbation can be reduced to zero (the theta condition) by an appropriate choice of solvent and temperature.

In unperturbed random coils, the distribution of end-to-end distances for long chains is Gaussian:

$$F(x_1, x_2, x_3)dx_1\,dx_2\,dx_3 = \left(\frac{3}{2\pi \langle r^2 \rangle} \right)^{3/2} \exp - \frac{3(x_1^2 + x_2^2 + x_3^2)}{2\langle r^2 \rangle} dx_1\,dx_2\,dx_3 \quad (2.1)$$

in which $F(x_1, x_2, x_3)dx_1\,dx_2\,dx_3$ is the fraction of chains with end-to-end coordinate differences in the range

$$x_1, x_1 + dx_1; \quad x_2, x_2 + dx_2; \quad x_3 + dx_3,$$

and $\langle r^2 \rangle$ is the mean square end-to-end distance:

$$\langle r^2 \rangle = \int\limits_{-\infty}^{\infty} \int \int (x_1^2 + x_2^2 + x_3^2)\, F(x_1, x_2, x_3)\, dx_1\,dx_2\,dx_3.$$

For random coils, $\langle r^2 \rangle$ is directly proportional to the contour length. If n is the number of main chain atoms in the chain, $\langle r^2 \rangle = an$. The parameter a is relatively insensitive to environment (21), and has been calculated for a number of polymers from strictly intramolecular considerations using the rotational isomeric model (22). The root-mean-square distance of segments from the center of gravity of the coil is called the radius of gyration S. The quantity S^3 is an approximate measure of the pervaded volume of the coil. For Gaussian coils,

$$S^2 = \langle r^2 \rangle / 6. \tag{2.2}$$

For dilute solutions in good solvents the net excluded volume is positive, and coil dimensions are expanded beyond their unperturbed values. The expansion ratio α,

$$\alpha^2 = \langle r^2 \rangle / \langle r^2 \rangle_\theta, \tag{2.3}$$

increases with chain length (19). A repulsive potential exists between chains in good solvents because of volume exclusion, so the distribution of molecular centers of gravity in dilute solutions is probably not quite random.

It is generally supposed that coil dimensions in good solvents decrease with increasing concentration, the intramolecular effects of excluded volume on dimensions gradually being offset by a forced increase in intermolecular contacts. How rapid the contraction with increasing concentration is not known with certainty, since unambiguous methods for determining coil dimensions in the moderate concentration range are not available. The calculations of Fixman and Peterson (23) and Yamakawa (24) suggest that dimensions in good solvents contract rather rapidly through the dilute range until the coils begin to overlap ($c[\eta] \approx 1$). Above this range the local density of polymer segments becomes uniform, and the molecular centers again become randomly distributed in the solution. According to Fixman and Peterson (23), the coils continue a gradual concentration at higher concentrations, the expansion factor $\alpha^2 - 1$ eventually becoming inversely proportional to polymer concentration. The actual magnitude of $\alpha^2 - 1$ in the concentrated region of in undiluted polymers is not given by the theory. Computer simulations of volume-excluding polymers (25, 26) support the idea of coil concentration with increasing concentration in the dilute range.

In Flory's theory of the excluded volume (27), the chains in undiluted polymer systems assume their unperturbed dimensions. The expansion factor in solutions is governed by the parameter $(\frac{1}{2} - \chi)/v$, v being the molar volume of solvent and χ the segment-solvent interaction (regular solution) parameter. In undiluted polymers, the solvent for any molecule is simply other polymer molecules. If it is assumed that the excluded volume term in the thermodynamic theory of concentrated systems can be applied directly to the determination of coil dimensions, then χ is automatically zero but v is very large, reducing the expansion to zero.

Taken literally, however, this viewpoint leads to rather strange variations of dimensions at intermediate concentrations. Thus (27),

$$\alpha^5 - \alpha^3 = \frac{K}{v}\left(\frac{1}{2} - \chi\right) \tag{2.4}$$

in which K is a concentration independent parameter for the given polymer-solvent system. From the standpoint of any polymer molecule in the system the surroundings consist of a mixture of solvent and other polymer molecules. The mean interaction coefficient with this two-component environment might reasonably be taken to be $\chi_0(1 - \varphi)$, where χ_0 is the interaction coefficient with pure solvent as the surroundings and φ is the volume fraction of polymer. The molar volume of the environment is

$$v = 1/(\varphi/v_p + (1 - \varphi)/v_s) \approx v_s/(1 - \varphi),$$

v_s and v_p being the molar volumes of pure solvent and polymer, and $v_p \gg v_s$. Thus:

$$\alpha^5 - \alpha^3 = \frac{K}{v_s}\left[\frac{1}{2} - \chi_0(1 - \varphi)\right](1 - \varphi). \tag{2.5}$$

Accordingly, coil dimensions in typical good solvents ($\chi_0 = 0.4$) should expand at intermediate concentrations, pass through a maximum near $\varphi = 0.35$ and finally subside slowly toward the theta dimensions at very high concentrations. Even without the modification of χ, the molar volume effect makes $\alpha^5 - \alpha^3$ proportional to $1 - \varphi$, which suggests that coil dimensions are still only partially contracted even out to rather high concentrations.

2.2.2. Evidence on Chain Dimensions from Physical Measurements

Krigbaum and Godwin (28) have estimated the end-to-end distance of individual chains in undiluted polystyrene, using low angle X-ray scattering. Polystyrene molecules ($\bar{M}_n = 87000$) which had been tagged at both ends by silver atoms were dispersed at a concentration of 5% in untagged polystyrene. Analysis of the excess scattering yielded $\langle r^2 \rangle^{1/2} = 269$ Å. The value in θ-solvents for this molecular weight is 235 Å.

Very recently, results from low angle neutron scattering with undiluted mixtures of deuterated and ordinary polymers have begun to appear. In these cases all units along the chain contribute to the excess scattering, and the radius of gyration is obtained. Cotton *et al.* (29) have measured the dimensions of a rather low molecular weight ($M = 8000$) deuterated polystyrene in ordinary polystyrene, obtaining $S = 28$–30 Å ($S_\theta = 26$ Å). Ballard *et al.* (*30*) dispersed ordinary poly-styrene ($\bar{M}_w = 97100$; $\bar{M}_w/\bar{M}_n = 1.06$) to a concentration of 5% in a matrix of 93% deuterated polystyrene ($\bar{M}_w = 99200$; $\bar{M}_w/\bar{M}_n = 1.1$), obtaining $S = 90$ Å ($S_\theta = 84$ Å).

Kirste *et al.* (*31*) dispersed ordinary polymethyl methycrylate ($\bar{M}_w = 250000$; $\bar{M}_w/\bar{M}_n = 1.36$) to a concentration of 1.2% in completely deuterated polymethyl methacrylate ($\bar{M}_w = 250000$; $\bar{M}_w/\bar{M}_n = 2.1$). They report $S = 126$ Å, compared to $S_\theta = 110$ Å, and $S = 170$ Å in dilute solutions of an athermal solvent. The same workers have recently studied the concentration dependence of the excess scattering (constructing in effect a Zimm plot) and obtain $S = 121$ Å at infinite dilution in the deuterated matrix, with $A_2 \approx 0$ and a molecular weight (by neutron scattering measurements alone) $\bar{M}_w = 220000$ (*32*). Moreover, they have deter-mined dimensions for a series of molecular weights ($\bar{M}_w$ ranging from 80000 to 1000000) and find $S \propto M^{0.49}$, although within their limits of error the θ-solvent exponent of 0.5 fits also.

The results of the scattering experiments are shown in Table 2.1. They leave little room for doubt as to the essential correctness of Flory's deduction about chain dimensions in undiluted polymers.

Table 2.1. Chain dimensions in undiluted amorphous polymers obtained by scattering experiments

Polymer	$\bar{M}_w$	$\bar{M}_w/\bar{M}_n$	S Å	S_θ Å	Method
Polystyrene	87000[a]	unreported but probably narrow	269[b]	235[b]	X-ray scattering from chains with Ag-tagged ends (8)
Polystrene	8000	unreported	28–30	26	Neutron scattering from deuterated chains (10)
Polystrene	97100	1.06	90	84	Neutron scattering using deuterated substrate (10a)
Polymethyl methacrylate	250000	1.36	130	110	Neutron scattering using deuterated substrate (11)

[a] Only $\bar{M}_n = 87000$ was reported.
[b] $\langle r^2 \rangle^{1/2}$ rather than S was measured.

Bueche *et al.* (*33*) determined chain dimensions indirectly, through mea-surements of the diffusion coefficient of C^{13}-tagged polymers in concentrated solutions and melts. The self-diffusion coefficient is related to the molar frictional coefficient $N_a n \zeta_0$ through the Einstein equation:

$$D = \frac{RT}{N_a n \zeta_0} \tag{2.6}$$

in which N_a is Avogadro's number, n is the number of main chain atoms per molecule, and ζ_0 is the frictional coefficient per main chain atom. The viscosity of concentrated solutions or melts in the free-draining Bueche-Rouse model (7) is:

$$\eta = \frac{N_a n \zeta_0 c \langle r^2 \rangle}{36\,M}. \tag{2.7}$$

Elimination of the frictional coefficient between Eqs. (2.6) and (2.7) yields

$$\frac{D\eta}{cRT} = \frac{\langle r^2 \rangle}{36\,M}. \tag{2.8}$$

For polystyrene fractions in diethyl phthalate solution ($30000 < M < 200000$; $0.3 < c < 0.6$ gm/ml) the group on the left had an average value of 1.6×10^{-18} ($\pm 50\%$). In dilute solution $\langle r^2 \rangle_\theta / 36\,M$ is 1.27×10^{-18} for polystyrene (21). No systematic variations with concentration, molecular weight or temperature were apparent, the scatter of the data being mainly attributable to the experimental difficulties of the diffusion measurements. The value of $D\eta/cRT$ for an undiluted tagged fraction of poly(n-butyl acrylate) in pure polymer was found to be 2.8×10^{-18}. The value of $\langle r^2 \rangle_\theta / 36\,M$ is 1.6×10^{-18}, based on dilute solution data for other acrylate polymers (34). Thus, transport behavior, like the scattering experiments, supports random coil configuration in concentrated systems, with *perhaps* some small expansion beyond θ-dimensions.

Semi-quantitative arguments in favor of the random coil have been made on the basis of mechanical behavior. For example, Bueche and co-workers (35) estimated molecular dimensions from an analysis of fracture envelopes in rubbery polymers. They suggested that the elongation at break corresponds to a maximum extension of strands between cross-links or entanglement junctions. The end-to-end distance for strands in the undeformed state $\langle r^2 \rangle^{1/2}$ was obtained from the molecular weight between junction points, estimated from other experiments, and the macroscopic deformation at break. Four different polymers were examined; the values obtained for $\langle r^2 \rangle$ in each case fell within 30% of corresponding θ-solvent values. Later work by Smith and Frederick (36) on a wide variety of vulcanizates showed that Bueche's equations indeed gave correct orders of magnitude for $\langle r^2 \rangle$, but the residual variations among polymers did not correlate very well with independently measured θ-dimensions.

Hoffman (37) has offered a variety of circumstantial evidence supporting the random coil model. In A–B block copolymers of styrene and butadiene, for instance, the characteristic dimension of the dispersed phase particles depends on the molecular weight of blocks in the dispersed phase according to:

$$D = KM^{0.58}. \tag{2.9}$$

The exponent would be 0.33 for collapsed coils, 0.5 for random coils, and 1.0 for fully extended chains. The observed values of D lie within a factor of three of the dilute solution values for $\langle r^2 \rangle^{1/2}$ of molecules with the same molecular weight as the blocks.

2.2.3. Collapsed Coil Theories

Other chain conformations have been suggested for concentrated systems. Maron and co-workers (38) have proposed that the molecules collapse into increasingly compact coils as concentration increases, with relatively little inter-penetration and segment mixing even at very high concentrations. They base this conclusion on solution viscosities (0–20% polymer) in several polymer systems, analyzed with an equation which had previously been found useful in correlating viscosity in suspensions of spherical particles. In the work on suspensions, the viscosity had been found to be principally a function of the volume fraction of solvent, expressed as $1 - \varepsilon \Phi$. In the case of suspensions Φ is the volume fraction of particles and ε is a factor interpreted in terms of the maximum packing density of spheres. In suspensions, ε is a constant, equal to 1.35. In polymer solutions the empirically deduced value of ε is a decreasing function of concentration with the form:

$$\frac{\varepsilon}{\varepsilon_0} = \frac{1}{1 + (\varepsilon_0 - \varepsilon_\infty)\varphi} \tag{2.10}$$

where φ is the volume fraction of polymer in the system, ε_0 is $[\eta]\varrho/2$, ε_∞ is approximately 4.0, ϱ is the density of undiluted polymer, and $[\eta]$ is its intrinsic viscosity in the solvent in question. Particle-like flow properties are indeed observed in the dilute to moderately concentrated range for polymer solutions (see Section 5). Moreover the idea of coil contraction with increasing concentration, due to cancellation of excluded volume in good solvents, is quite a reasonable idea, as noted earlier. However, the authors go on to infer from this result that coil contraction continues unabated into the regime of concentrated solutions and melts, with relatively little intermolecular contact except at the coil periphery at all concentration levels.

Maron has incorporated this collapsing coil picture into a thermodynamic theory of polymer solutions (39). The free energy of mixing for n_1 moles of solvent and n_2 moles of polymer at constant volume turns out to be the Flory-Huggins expression with two modifications, an additional term containing the contraction ratio $\varepsilon/\varepsilon_0$ and a somewhat different handling of the interaction coefficient χ, the latter having nothing directly to do with coil contraction:

$$\Delta G_M = RT\left[n_1 \ln(1 - \varphi) + n_2 \ln \varphi + n_2 \ln \frac{\varepsilon}{\varepsilon_0} + \chi n_1 \varphi\right]. \tag{2.11}$$

This equation was then applied to a variety of experimental data on solvent activity as related to vapor pressure (40), osmotic pressure in dilute and concentrated solutions (41), and light scattering (42).

Both modifications affect the analysis of dilute solution behavior, and it is difficult to judge how much the $\varepsilon/\varepsilon_0$ term is actually needed. In any case, as the authors themselves point out (41), the $\varepsilon/\varepsilon_0$ term makes an entirely negligible contribution to solvent activity in concentrated solutions. For example, simple calculations yield a contribution of approximately 1% in a 10% solution of natural rubber in benzene at 30° C ($M = 500000$, $[\eta]\varrho = 250$, $\chi = 0.4$). It is therefore clear that thermodynamic measurements can furnish no evidence for or against continued collapse in concentrated solutions.

Aharoni has stated that the observed rates of crystallization in polymers are inconsistent with the times required for random-coil molecules to separate themselves from the melt, and claims this as support for the collapsed coil model (43). No numerical comparisons are given, and it is difficult therefore to judge the basis for his assertion.

2.2.4. Evidence on Chain Dimensions from Chemical Properties

Vollmert and Stutz (44) suggest that they have found support for a partially collapsed coil in concentrated solutions from an extensive study of the reactions between functional groups located on different polymer molecules. They prepared copolymers of n-butyl acrylate, one with 5% alcohol functions, the other with 5% acid chloride functions. Equal amounts of these copolymers were dissolved in benzene and allowed to mix for several hours at room temperature. Reaction to form ester cross-linkages was then brought about by heating to 70° C. The extent of reaction was much lower than when the copolymer containing alcohol functions was cross-linked with equivalent amounts of a monomeric dichloride. To separate out segregation effects due to thermodynamic incompatibility, gelation experiments (on the same acrylic substrate but with an alcohol-isocyanate curing system this time) were conducted over a range of copolymer composition and extrapolated to 0% reactive functions. Segregation effects are rather clearly evident even with relatively low copolymer contents. The extrapolation is therefore a difficult and sensitive one. The values obtained by the authors are given in Table 2.2.

Table 2.2. Extents of reaction in acetone between alcohol and isocyanate functions on different poly(n-butyl acrylate) molecules, extrapolated to 0% reactive functions

Polymer concentration (%)	Extent of reaction (%)
10	2
15	5
20	9
25	13.5
30	17

The authors propose that these extents of reaction reflect the degree of coil overlap at each polymer concentration. They conclude that each molecule in concentrated solution has a segregated core, and that the intermolecular contact occurs only near the periphery. As the concentration increases the overall molecular size must shrink correspondingly. [We note in passing that fluorescence and fluorescence quenching by groups attached to different polymer molecules appears to be a promising alternative method for analyzing segregation in concentrated systems (45), although no results applicable to questions of coil dimensions have appeared as yet.]

It seems impossible to reconcile such collapsed coil models with a wealth of other observations on chemical reactivity in amorphous polymers. For example, any significant collapse of molecular size in concentrated systems should cause deviations from the predictions of random cross-linking theory. The theory assumes that all repeating units in a molecule have equal probabilities of undergoing cross-link formation. Only intermolecular cross-links can be effective in developing a gel structure. Collapsed coils would result in appreciable amounts of intramolecular cross-linking, and the intermolecular cross-linking density would be proportional to the molecular surface area, $M^{2/3}$, rather than total molecular weight M as in the case of intermingling coils. Gel point dosages R_g, corrected for chain scission and small amounts of end-linking, are shown in Table 2.3 for polystyrene films which were cross-linked by gamma radiation (46).

Table 2.3. Gel points in polystyrene cross-linked by gamma radiation

$\bar{M}_w$	R_g (megarads)	$R_g\bar{M}_w$	$R_g\bar{M}_w^{2/3}$
0.0562×10^6	381.0	2.1×10^7	5.6×10^5
0.082	256.0	2.1	4.9
0.186	105.0	1.9	3.4
0.329	58.4	1.9	2.8
0.362	63.6	2.3	3.2
0.915	22.8	2.1	2.2
1.87	12.0	2.2	1.8
1.90	11.9	2.3	1.8
3.37	6.33	2.1	1.4

The total number of cross-links per gram is directly proportional to radiation dose. Either $R_g\bar{M}_w$ or $R_g(\bar{M}_w)^{2/3}$ should be independent of initial molecular weight, depending on whether all repeating units or only those near the surface of the coils have equal probabilities for intermolecular cross-link formation. Complete coil collapse is clearly ruled out by the results. Moreover, detailed calculations (46) show that a trend in the product $R_g\bar{M}_w$ due to systematic differences in intramolecular cross-linking, should be observed even for only partially collapsed random coils. If the coil radii in bulk polystyrene were reduced by even as much as a factor of two from their θ-solvent values, a different of 15% between the extremes of molecular weight would be expected. No such trend is observed.

The frequency of intramolecular cross-linking can be estimated from the material balance on H_2 in linear polyethylene irradiation. The total number of cross-links (obtained from gel point determination) plus the number of unsaturated links formed (obtained from infrared measurements) minus the number of chain scission reactions (obtained from the limiting solubility at high radiation doses) should be equal to the number of H_2 molecules evolved, assuming all cross-links are intermolecular. There does indeed appear to be a discrepancy in the material balance for irradiation in the melt state, implying some intramolecular linking, although there is some dispute about its actual magnitude (47, 48). In any case, the discrepancy appears to be independent of molecular weight, an unlikely circumstance for collapsed coils since intramolecular cross-linking would surely vary with molecular weight for that configuration.

More quantitative chemical evidence for random coil configuration comes from cyclization equilibria in chain molecules (49). According to the random coil model there must be a very definite relationship among the concentrations of x-mer rings in an equilibrated system, since the cyclization equilibrium constant K_x should depend on configurational entropy and therefore on equilibrium chain and ring dimensions. Values of $\langle r^2 \rangle / M$ deduced from experimental values on K_x for polydimethylsiloxane, both in bulk and in concentrated solution, agree very well with unperturbed dimensions deduced from dilute solution measurements(49).

2.2.5. Extended Configuration Theories

Kargin and co-workers (50) have published extensively on supermolecular structures in crystalline polymers. Based principally on the surface morphologies observed by electron microscopy, Kargin has proposed that even in amorphous polymers the chains are aligned in loose bundle-like structures which persist over distances which are large compared to molecular dimensions. Direct supporting evidence for such expanded chain configurations has not been forthcoming however.

Tager and co-workers (51) have invoked bundle structures to explain correlations between the viscosities of concentrated polymer solutions and the thermodynamic interactions between polymer and solvent. They note, for example, that solutions of polystyrene in decalin (a poor solvent) have higher viscosities than in ethyl benzene (a good solvent) at the same concentration, and quote a number of other examples. Such results are attributed to the ability of good solvents to break up the bundle structure; the bundles presumably persist in poor solvents and give rise to a higher viscosity. It seems possible that such behavior could also be explained, at least in part, by the effects of solvent on free volume (see Section 5). Berry and Fox have found, for example, that concentrated solution data on polyvinyl acetate in solvents of quite different thermodynamic interaction could be reduced satisfactorily by free volume considerations alone (16). Differences due to solvent which remain after correction for free volume effects might reasonably be attributed to association, although not necessarily to bundle structures.

Tager and Dreval (52) have also noted that the shear rate dependence of viscosity in concentrated solutions varies with the thermodynamic character of the solvent. In appropriately reduced form, the onset of shear rate dependence appears to be independent of solvent character (53), but at high shear rates the viscosity levels off in poor solvents while continuing to decrease in good solvents (52). The bundle picture is again invoked, the argument apparently being that some bundles are so strong in poor solvents that they prevent the viscosity from falling beyond on a certain point. Viscosity anomalies have been reported by others for solutions of moderate concentration near the theta temperature (54, 55). However, in these cases the anomalies were found in company with critical point concentration fluctuations (54) or with the actual separation of a second phase (55).

2.3. Short-Range Organization

Local structural features have been postulated for amorphous polymer systems, based on the asymmetry of chain-like molecules. Flory (56) has shown that molecular asymmetry in itself is no barrier to a dense random packing of the chains are sufficiently flexible. Robertson (57) suggests, however, that some degree of local alignment is required simply to accomodate linearly connected sequences in the rather limited space available. Unfortunately, calculations of local cooperative effects are extremely difficult and sensitive to specific assumptions about available packing arrangements.

Small nodular regions ($D \approx 50$–150 Å) have been observed by electron microscopy in thin films of amorphous but crystallizable polymers such as polyethylene terephthate (58) and polycarbonate (59). Scattering regions of approximately the same size have also been detected by low-angle X-ray diffraction (60). Yeh has observed similar structures in atactic polystyrene (61). The nodule diameter in the latter polymer is approximately 30 Å, corresponding to a grouping of about 300 repeating units. Yeh also finds evidence for strongly diffracting regions by dark field examination of the amorphous halo (62). He has proposed that there are small regions even in non-crystallizable polymers where the chain segments are highly ordered, and suggests that the properties of amorphous polymers may in fact be strongly influenced by such structures (63). The packing in these regions cannot be much different from the remainder of the system, however, since even in polycarbonates the density difference can be no more than about 2% (39).

Flory has recently summarized the experimental evidence pertaining to local correlation and their effects on chain dimensions (49). There is experimental support for local alignment from optical properties such as stress-optical coefficients in networks (both unswelled and swelled in solvents of varying asymmetry), and from the depolarization of scattered light in the undiluted state and at infinite dilution. The results for polymers however, turn out to be not greatly different from those for asymmetric small molecule liquids. The effect of these correlations on thermodynamic and mechanical properties appears to be very small in non-polar polymers. Molecular asymmetry of the swelling solvent has no discernable effect on the modulus of polyethylene (64) and polyisoprene (65) networks, after the usual corrections for variations with swelling ratio are made. The temperature coefficient for chain dimensions $d\ln\langle r^2\rangle/dT$ obtained from modulus measurements on networks agrees very well with values from dilute solution measurements (49). The entropy of melting and the activity of solvent in concentrated solutions appear to be understandable without postulating special polymeric association or alignment effects (66).

Thus, although some degree of local organization may indeed occur in amorphous systems, and may even have some effect on the mechanical properties of polymers in the glassy state, the influence on the mechanical properties of melts, concentrated solutions and networks appears to be negligible.

2.4. Structural Aspects of Flow and Flow History

Molecular theories of flow behavior are applied on the assumption that the macroscopic velocity field can be considered to apply without modification right down to the molecular scale. In continuum theories the components of relative velocity in an arbitrarily small neighborhood of any material point are taken to be linear functions of the spatial coordinates measured from that point, i.e., the flow is assumed to be locally homogeneous. The local velocity field is calculated from the macroscopic velocity field. This property of local homogeneity of flow is an obvious prerequisite for any meaningful macroscopic analysis, and perhaps the fact that analyses are at all successful and that flow properties can be determined which are independent of apparatus geometry constitutes a fair test of the assumption.

Certain systems may violate the assumption. Locally inhomogeneous deformation has been observed in thin films of atactic polystyrene below the glass temperature (67). Melts formed from polyvinyl chloride particles retain a particulate appearance in some cases even after extrusion (68). Collins and Krier (68a) have shown that molecular weight changes during flow have little influence on apparent viscosity in polyvinyl chloride below 190–200° C. At higher temperatures the viscosity responds immediately when molecular weight changes. The clear implication is a change in the local flow pattern near 200° C from particulate flow to molecular flow.

Schreiber and co-workers have noted very persistent history effects in linear polyethylenes (69). Fractions which have been crystallized from dilute solution required times of the order of hours in the melt state at 190° C in order to attain a constant die swell behavior upon subsequent extrusion. The viscosity on the other hand reached its ultimate value almost immediately. The authors concluded from this result that different types of molecular interactions were responsible for elastic and viscous response. However, other less specific explanations might also suffice, since apparent viscosity might be relatively intensitive to the presence of incompletely healed domain surfaces, while die swell, requiring a coordinated motion of the entire extrudate, might be affected by planes of weakness. It would be interesting to repeat such experiments in a plate-cone instrument where elastic response can be measured more directly.

Ballman and co-workers have used carbon particles to determine flow patterns for polystyrene melts in plate-cone and capillary viscometers (70). Complex patterns, rather than the simple flow expected, were observed for high molecular weight samples. These may have been caused, however, by differences in viscosity between adjacent layers of pure melt and melt with suspended particles.

Mooney earlier had suggested a kind of ball-bearing flow for unvulcanized rubber in plate-cone viscometers, based on an observed rapid rate of diffusion of dye in the direction of the velocity gradient (71, 72). Flow instabilities of various kinds, melt fracture in capillary flow and spontaneous extrusion from the gap in plate-cone flows, are known to be associated with high shear rates in viscoelastic materials. Judged from the combination of high shear rates and viscosities in the Mooney experiments, it seems likely in retrospect that the rapid mixing behavior was obtained under conditions of unstable flow. Most recent studies of rheological properties seek to avoid conditions which lead to such flows. Locally homogeneous deformations apparently can be achieved in systems which are initially uniform if the deformations or deformation rates are sufficiently small. The onset of flow instabilities appears to depend on flow geometry, and occur at much higher shear rates for capillary flows than plate-cone flows.

A related question concerns the effect of prior flow history on the physical structure of the system. Flow would be expected to produce a partial alignment of the chains. This orientation can be frozen in by rapid cooling but should be lost through configurational relaxation when the sample is reheated to the melt state. Wissbrun has noted a reduction in both viscosity and die swell in branched poly-methylene oxide after an initial extrusion (73). Prolonged annealing restored only part of the original behavior, and a more complete recovery of properties was only achieved by dissolving and re-precipitating the polymer.

Schreiber and co-workers have found some indication of molecular frac-tionation during capillary flow, with smaller molecules migrating preferentially towards the wall and larger molecules toward the center line (74). This conclusion has been disputed recently by Porter and co-workers however (75).

Numerous experiments have shown that recovery from previous flow history is smooth and relatively rapid in polymer melts and solutions. Recovery is especially rapid for systems of narrow molecular weight distribution. Agreement between different investigators, and agreement in rheological information obtained in different flow geometries also suggest that effects of prior history, although crucial in solid state properties, is not a significant problem in most rheological studies.

2.5. Summary

Both direct measurements and a wealth of compelling indirect evidence support the random coil, *perhaps* slightly expanded beyond θ-dimensions, as the correct description of large-scale chain configuration in amorphous systems of flexible, relatively non-polar polymers. It is certain that neither greatly contracted nor highly extended configurations can be reconciled with these data. Evidence does exist for short-range intermolecular correlations which are probably related to molecular asymmetry. The degree of orientation correlation is similar in extent to that in systems of asymmetric small molecules. Its effect on the mechanical and thermodynamic properties of polymer systems appears to be negligible in the cases so far investigated. Judged by agreement among investigators using different samples and flow geometries, the effects of preparation history on flow properties is negligible in narrow distribution linear polymers, and the flow pattern itself is locally homogeneous in the absence of various types of viscoelastic instability.

3. Material Properties of Viscoelastic Liquids

This section summarizes results of the phenomenological theory of viscoelasticity as they apply to homogeneous polymer liquids. The theory of incompressible simple fluids (76, 77) is based on a very general set of ideas about the nature of mechanical response. According to this theory the flow-induced stress at any point in a substance at time t depends only on the deformations experienced by material in an arbitrarily small neighborhood of that point in all times prior to t. The relationship between stress at the current time and deformation history is the constitutive equation for the substance.

The stress tensor describes the forces transmitted to an element of material through its contacts with adjacent elements (78). Traction is the force per unit area acting outwardly on the material adjacent to a material plane, and transmitted through its contact with material across the plane. If the components of traction are known for any set of three planes passing through a point, the traction across any plane through the point can be calculated. The stress at a material point is determined by an assembly of nine components of traction, three for each plane. If the orientations of the three planes are chosen to be normal to the coordinate directions of a rectangular Cartesian coordinate system, the Cartesian components of the stress are obtained:

$$\boldsymbol{p} = \begin{bmatrix} p_{11} & p_{12} & p_{13} \\ p_{21} & p_{22} & p_{23} \\ p_{31} & p_{32} & p_{33} \end{bmatrix}. \tag{3.1}$$

The stress tensor is symmetric, so $p_{ij} = p_{ji}$. Furthermore, addition of an isotropic pressure has no effect on the physical properties of incompressible substances, so only differences among the normal components of stress are governed by the constitutive equation. Thus, there are at most only five independent terms of rheological significance in $\boldsymbol{p}$, three shear components ($p_{12}, p_{13},$ and $p_{23},$ for example) and two differences in the normal components ($p_{11} - p_{22}$ and $p_{22} - p_{33},$

for example). In incompressible simple fluids these attributes of the stress are governed only by the history of the relative deformation gradient. The property of fading memory requires that recent strains relative to the current configuration be weighted more heavily in determining the current stress than strains in the more distant past.

The response of simple fluids to certain classes of deformation history can be analyzed. That is, a limited number of material functions can be identified which contain all the information necessary to describe the behavior of a substance in any member of that class of deformations. Examples are the viscometric or steady shear flows which require, at most, three independent functions of the shear rate (79), and linear viscoelastic behavior (80, 81) which requires only a single function, in this case a relaxation function. The functions themselves must be determined experimentally for each substance.

The utility of the simple fluid theory lies in the plausibility and generality of its assumptions about how materials behave and in the exactness with which its conclusions are worked out. In particular, one is inclined to believe, as a working hypothesis and in the absence of contradictory evidence, that the theory is general enough to encompass the behavior of homogeneous polymeric liquids. On this basis the role of molecular theories is a complementary one, to provide forms for the material functions and to account for their systematic change with molecular structure and temperature.

Most rheological data on polymer liquids of known structure has been obtained in simple shearing deformations. The velocity field for homogeneous simple shear in rectangular Cartesian coordinates may be expressed:

$$v_1 = \dot{\gamma}(t)x_2$$

$$v_2 = 0 \tag{3.2}$$

$$v_3 = 0$$

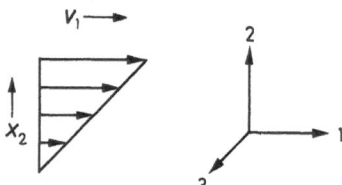

in which the coordinates $x_i(i=1, 2, 3)$ are measured from some suitably chosen fixed origin, and in which 1 denotes the direction flow, 2 is the direction normal to the shearing planes, and 3 is the third or neutral direction. The shear rate, $\dot{\gamma}(t)$, characterizes the flow field; the current stress depends on the history of the shear deformation $\gamma(t)$:

$$\gamma(t) = \gamma(0) + \int_0^t \dot{\gamma}(s)\,ds \tag{3.3}$$

i.e., the values of the shear deformation at all past times.

Symmetry arguments suffice to show that p_{13} and p_{23} are identically zero for this deformation field (79):

$$\mathbf{p}(t) = \begin{bmatrix} p_{11} & p_{12} & 0 \\ p_{21} & p_{22} & 0 \\ 0 & 0 & p_{33} \end{bmatrix}. \tag{3.4}$$

Thus, the effects of deformation on the stress tensor reduce to the effect of the values of $\gamma(t-s)$ from $s=0$ to $s=\infty$ on the values of $p_{12}\,(=p_{21})$, $p_{11}-p_{22}$, and $p_{22}-p_{33}$ at time t. The shear stress will be denoted by σ in all further discussions.

For steady shear flow, the shear rate $\dot{\gamma}$ is constant for all past time. Since deformation history now depends only on the parameter $\dot{\gamma}$, the stress components become functions of $\dot{\gamma}$ alone:

$$\sigma = \eta(\dot{\gamma})\,\dot{\gamma} \tag{3.5}$$

$$p_{11} - p_{22} = N_1 = \Psi_1(\dot{\gamma})\,\dot{\gamma}^2 \tag{3.6}$$

$$p_{22} - p_{33} = N_2 = \Psi_2(\dot{\gamma})\,\dot{\gamma}^2. \tag{3.7}$$

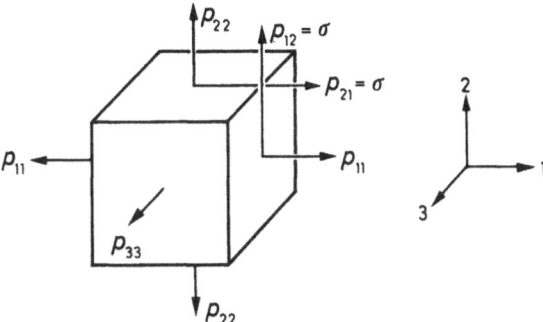

The symbols N_1 and N_2 denote the normal stress functions in steady state shear flow. Symmetry arguments show that the viscosity function $\eta(\dot{\gamma})$ and the first and second normal stress coefficients $\Psi_1(\dot{\gamma})$ and $\Psi_2(\dot{\gamma})$ are even functions of $\dot{\gamma}$. In the limit of zero shear rate $\eta(\dot{\gamma})$ and $\Psi_1(\dot{\gamma})$ become constants, termed the zero-shear viscosity η_0 and the zero-shear normal stress coefficient $\Psi_1(0)$ respectively.

Of major interest in this review are $\eta(\dot{\gamma})$ and $\Psi_1(0)$ for which a large quantity of data has now been accumulated on well-characterized polymers. Some limited information is also available on the shear rate dependence of Ψ_1. The second normal stress function has proved to be rather difficult to measure; N_2 appears to be negative and somewhat smaller in magnitude than N_1 (82).

Viscoelastic behavior is classified as linear or non-linear according to the manner by which the stress depends upon the imposed deformation history (80). Insteady shear flows, for example, the shear rate dependence of viscosity and the normal stress functions are non-linear properties. Linear viscoelastic behavior is obtained for simple fluids if the deformation is sufficiently small for all past times (infinitesimal deformations) or if it is imposed sufficiently slowly (infinitesimal rate of deformation) (80, 83). In shear flow under these circumstances, the normal stress differences are small compared to the shear stress, and the expression for the shear stress reduces to a statement of the Boltzmann superposition principle (15, 81):

$$\sigma(t) = \int_0^\infty G(s)\, \dot{\gamma}(t-s)\, ds .\tag{3.8}$$

The shear stress relaxation modulus of the fluid, $G(t)$, is a monotonically decreasing function of time, with $G(\infty) = 0$. If the fluid initially at rest is given a small shear deformation γ_0 at $t = 0$, the shear stress at later times becomes simply:

$$\sigma(t) = \gamma_0\, G(t) .\tag{3.9}$$

The value $G(0)$ is called the instantaneous modulus, and will be denoted by G^0.

It turns out that stress relaxation following a simple shear deformation is seldom employed experimentally. A more common technique is to measure the steady state response to small sinusoidal deformations as a function of angular frequency ω. The dynamic storage modulus $G'(\omega)$ and loss modulus $G''(\omega)$ in small sinusoidal deformations are related to $G(t)$:

$$G'(\omega) = \omega \int_0^\infty G(t) \sin \omega t\, dt ,\tag{3.10}$$

$$G''(\omega) = \omega \int_0^\infty G(t) \cos \omega t\, dt .\tag{3.11}$$

The instantaneous modulus is given by:

$$G^0 = \frac{2}{\pi} \int_{-\infty}^\infty G''(\omega)\, d\ln\omega .\tag{3.12}$$

In creep and creep recovery experiments, the stress is imposed and the deformation is observed. Equation (3.8) can be inverted to describe those cases:

$$\gamma(t) = \int_0^\infty J(s)\,\dot\sigma(t-s)\,ds.\tag{3.13}$$

The shear creep compliance, $J(t)$, is related to the relaxation modulus through:

$$t = \int_0^t G(t-s)\,J(s)\,ds.\tag{3.14}$$

If a sufficiently small constant shearing stress σ_0 is imposed on the material at $t=0$, the subsequent deformation is given by

$$\gamma(t) = J(t)\,\sigma_0.\tag{3.15}$$

At long times, the deformation rate becomes constant, and the creep function approaches a straight line:

$$J(t) = J_e^0 + t/\eta_0\tag{3.16}$$

in which η_0 is the zero shear viscosity and J_e^0 is the steady state recoverable shear compliance, or more simply the steady state compliance. If the stress is removed at a time t_0, the liquid recoils and a portion of the total deformation is eventually recovered. If t_0 lies in the region of steady state deformation, the total recoverable shear or elastic recoil γ_r is

$$\gamma_r = J_e^0\,\sigma_0.\tag{3.17}$$

Both J_e^0 and η_0 can be evaluated from the response in other linear viscoelastic experiments also.

$$\eta_0 = \int_0^\infty G(t)\,dt = \lim_{\omega\to 0}\frac{G''(\omega)}{\omega},\tag{3.18}$$

$$J_e^0 = \frac{1}{\eta_0^2}\int_0^\infty t\,G(t)\,dt = \frac{1}{\eta_0^2}\lim_{\omega\to 0}\frac{G'(\omega)}{\omega^2}.\tag{3.19}$$

Behavior in steady state shearing flows likewise provide values of these parameters (83):

$$\eta_0 = \lim_{\dot{\gamma} \to 0} \eta(\dot{\gamma}), \tag{3.20}$$

$$J_e^0 = \frac{1}{2\eta_0^2} \lim_{\dot{\gamma} \to 0} \Psi_1(\dot{\gamma}). \tag{3.21}$$

The factor of two in Eq. (3.21) was a matter of some dispute until recently, but it now seems well established (84).

The linear viscoelastic properties are often expressed in terms of an auxiliary function, the relaxation time distribution, $H(\tau)$; $H(\tau) \, d \ln \tau$ is the portion of the initial modulus contributed by processes with relaxation times in the range $\ln \tau, \ln \tau + d \ln \tau$:

$$G(t) = \int_{-\infty}^{\infty} H(\tau) \, e^{-t/\tau} \, d \ln \tau. \tag{3.22}$$

This alternative representation is entirely arbitrary of course, but its use is well established in the literature. It offers some advantage in discussing the connections between the macroscopic and microscopic descriptions of viscoelasticity. The parameters, η_0, J_e^0, and G^0 can be expressed in terms of the moments of $H(\tau)$:

$$G^0 = \int_{-\infty}^{\infty} H(\tau) \, d \ln \tau, \tag{3.23}$$

$$\eta_0 = \int_{-\infty}^{\infty} \tau \, H(\tau) \, d \ln \tau, \tag{3.24}$$

$$J_e^0 = \frac{\int_{-\infty}^{\infty} \tau^2 \, H(\tau) \, d \ln \tau}{\left[\int_{-\infty}^{\infty} \tau \, H(\tau) \, d \ln \tau \right]^2}. \tag{3.25}$$

Highly entangled systems, especially those of narrow molecular weight distribution, are characterized by a set of relaxations at long times (terminal relaxations) which are more or less isolated from the more rapid processes. The modulus associated with the terminal processes is called the plateau modulus G_N^0. Because η_0 and J_e^0 depend on weighted averages over $H(\tau)$, their values are controlled almost completely by the terminal processes. These experimental parameters can therefore be combined to define average relaxation times in the *terminal region*:

$$\tau_n \equiv \frac{\int\limits_{-\infty}^{\infty} \tau\, H(\tau)\, d\ln\tau}{\int\limits_{-\infty}^{\infty} H(\tau)\, d\ln\tau} = \eta_0 / G_N^0, \tag{3.26}$$

$$\tau_w \equiv \frac{\int\limits_{-\infty}^{\infty} \tau^2 H(\tau) d\ln\tau}{\int\limits_{-\infty}^{\infty} \tau\, H(\tau)\, d\ln\tau} = \eta_0 J_e^0, \tag{3.27}$$

and

$$\frac{\tau_w}{\tau_n} = J_e^0 G_N^0. \tag{3.28}$$

The mean times τ_n and τ_w will be called the number-average and weight-average relaxation times of the terminal region, and τ_w/τ_n can be regarded as a measure of the breadth of the terminal relaxation time distribution. It should be emphasized that these relationships are merely consequences of linear visco-elastic behavior and depend in no way on assumptions about molecular behavior. The observed relationships between properties such as η_0, J_e^0, and G_N^0 and molecular parameters provides the primary evidence for judging molecular theories of the long relaxation times in concentrated systems.

Other types of linear viscoelastic experiments may be used. Dynamic shear compliance measurements provide the storage and loss compliances $J'(\omega)$ and $J''(\omega)$. An equation analogous to Eq.(3.12) is available for determining the initial modulus from $J''(\omega)$:

$$G^0 = \left[\frac{2}{\pi} \int\limits_{-\infty}^{\infty} J''(\omega) d\ln\omega \right]^{-1}. \tag{3.29}$$

Stress build-up at the beginning of a steady shearing flow and stress decay at its termination can also be employed. The analysis of results is somewhat different than in the more usual experiments, but again, as long as the response obeys the rules of linear viscoelasticity, the same fundamental information is obtained. Flow birefringence measurements can also be used, subject to the additional assumption that the stress and optical anisotropy tensors are linearly related. A large body of experimental evidence supports this assumption (18). For example, the recoverable compliance can be obtained from the extinction angle χ in steady shearing flow:

$$J_e^0 = \frac{1}{\eta_0} \lim_{\dot{\gamma}\to 0} \left[\frac{\cot 2\chi}{\dot{\gamma}} \right]. \tag{3.30}$$

Finally, tensile deformations provide the same information as shear deformation as long as the incompressibility assumption is not violated. In this case, the tensile stress relaxation modulus $E(t)$ is directly related to the shear modulus: $E(t) = 3 G(t)$, and all other relationships follow accordingly.

The experimental methods for measuring linear-viscoelastic properties are reviewed elsewhere (15). Whether the behavior being measured does, in fact, lie in the linear viscoelastic region must be judged separately in each system. Generally, tests are made at different levels of imposed stress, strain, or strain rate to establish that the same material function is obtained. In polymer systems the response is usually spread over many more decades of time or frequency than can be covered in a single isothermal experiment. Typically, response is measured over a convenient range at several temperatures and the results combined to form a master curve according to the principle of time-temperature superposition. The methods for forming a master response curve are described elsewhere (15).

4. Molecular Models in Polymer Rheology

4.1. Introduction

A variety of molecular models have been employed to represent polymer chains and their interactions with surroundings. Equilibrium properties and certain very simple transport properties depend mainly on the radial distribution of segments about the center of gravity of the molecule. In one group of models the segments are distributed in space according to their equilibrium positions; fluctuations in segment density and the correlations in position between successive segments are ignored. The particle cloud model (27) has been used to calculate thermodynamic properties of random coils in dilute solution, such as the virial coefficients and the expansion of the coil due to excluded volume. A similar porous sphere model has been applied to viscosity and diffusion calculations (85). These models are not well adapted for dealing with dynamic behavior, since mechanical connection along the chain and the consequent coupling in the motions of neighboring segments are crucial in such properties.

4.2. Necklace Models

Necklace models represent the chain as a connected sequence of segments, preserving in some sense the correlation between the spatial relationships among segments and their positions along the chain contour. Simplified versions laid the basis for the kinetic theory of rubber elasticity and were used to evaluate configurational entropy in concentrated polymer solutions. A refined version, the rotational isomeric model, is used to calculate the equilibrium configurational properties of chains from local bond angle restrictions and the differences in energy among the rotational isomeric states (22).

The pearl necklace model also provides the conceptual basis for theories of chain dynamics and viscoelastic properties. As formulated by Kramers (86) and, somewhat differently, by Kirkwood and co-workers (87), the configuration is specified by the positions of n main chain atoms. The positions of each atom relative to its nearest neighbors along the chain are subject only to bond length and valence bond angle restrictions (excluded volume effects are neglected). For simplicity, all allowed configurations are considered to have the same energy. The configuration changes when chain elements move to new positions by bond rotation. These changes are opposed by the viscous resistance of the medium. As the change takes place, the effects are felt by other segments in the vicinity (but perhaps remote along the chain contour) through the motions induced in the fluid. Random collisions between chain elements and their surroundings provide the impetus for configurational change in accordance with the theory of Brownian motion (88). The large-scale motions of the chain, those of interest in flow properties, arise out of the myriad of such primitive local bond rotations.

Necklace models can in principle deal with chain dynamics for all characteristic frequencies which lie well below the collision rate in the liquid ($\sim 10^{12}\,\mathrm{sec}^{-1}$), and with response to external fluid motions of arbitrary magnitude. In practice, they are seldom used because of the following problem. Given the local configurational restrictions, the subsequent movement of adjacent chain elements must somehow be coordinated in order to satisfy the restrictions (preservation of valence bond angles and lengths throughout the motion) and yet at the same time not to require large portions of the chain to move each time a local rotation occurs. It is not known how to determine the types of locally coordinated motions and their relative frequencies from the chain structure. The relative importance of the various possible motions probably are influenced by both the viscosity of the medium and the internal activation barriers.

Attempts have been made to identify primitive motions from measurements of mechanical and dielectric relaxation (89) and to model the short time end of the relaxation spectrum (90). Methods have been developed recently for calculating the complete dynamical behavior of chains with idealized local structure (91, 92). An apparent internal chain viscosity has been observed at high frequencies in dilute polymer solutions which is proportional to solvent viscosity (93) and which presumably appears when the external driving frequency is comparable to the frequency of the primitive rotations (94, 95). The beginnings of an analysis of dynamics in the rotational isomeric model have been made (96). However, no general solution applicable for all frequency ranges has been found for chains with realistic local structure.

4.3. Spring-Bead Models

The bead-spring models are devices to circumvent the complications of the local motion problem and still obtain information on the large-scale configurational relaxations which control viscoelastic behavior. Their utility lies in the intuitive notion, borne out by calculations with specific models for the primitive motions (91, 92), that the large-scale chain motions are largely independent of the details of the primitive motions. Stated another way, the thermal motions of long flexible molecules can be characterized by a set of relaxation times. The nature of the short time end of the spectrum is very sensitive to the details of local structure through the latter's influence on the primitive motions. However, the relative spacings of the longest relaxation times is independent of the primitive motions. Their net effect appears only as a featureless scaling factor, representable in terms of a segmental mobility or its reciprocal, a segmental frictional coefficient ζ.

The Rouse-Bueche model (97, 98) replaces the real molecule of n main chain atoms by a mechanical chain of $N+1$ beads joined in sequence by N linear springs. The frictional interactions with the medium, which are distributed uniformly along the length of the real molecule to give a molecular frictional coefficient $n\zeta_0$, are concentrated at regular intervals in the beads. The frictional coefficient for each bead is taken to be $\dfrac{n}{N+1}\zeta_0$, so the model has the same total frictional coefficient as the real molecule. The restraining effects of the n/N main chain bonds which connect the sites of the beads are provided by linear springs of spring constant $3NkT/\langle r^2 \rangle$; $\langle r^2 \rangle$ is the equilibrium mean-square end-to-end distance of the entire molecule and kT is the Boltzmann factor. This expression for the force is appropriate for Gaussian subchains undergoing random thermal motion (98, 99), and it provides the model with the same equilibrium distribution of large scale configurations as the real molecule. The actual value of N is immaterial, provided that n/N is large enough for the Gaussian approximation to be valid for the sub-chains, and that N itself is large enough for the frictional sites to be distributed in a fairly uniform manner over the chain contour. Calculated properties which turn out to be independent of N for $N \gg 1$ should be independent of local details of structure. Those which depend on N are, by inference, sensitive to local structure and cannot be expected to be well represented by spring-bead models.

The configuration of any chain is specified by the $3(N+1)$ coordinates of its beads. In an ensemble of ν chains per unit volume, $\Psi \prod_{m=0}^{N} d\mathbf{r}_m$ is the probability of finding a chain with a specified set of bead coordinates $\mathbf{r}_m (m=0, 1, ..., N)$, where $\mathbf{r}_m$ is the position vector of bead m. Conservation of beads in the system requires the probability distribution function Ψ to satisfy the equation of continuity:

$$\frac{\partial \Psi}{\partial t} + \sum_{m=0}^{n} \tilde{\nabla}_m \cdot (\dot{\mathbf{r}}_m \Psi) = 0 \qquad (4.1)$$

in which $\dot{\mathbf{r}}_m$ is the velocity of bead m.

Inertial forces are neglected, and the chains are assumed to interact with the medium independently. The flow is assumed to be locally homogeneous, i.e., the velocity of the medium at each bead position is taken to be a linear function of the bead coordinates. A balance of the mean forces on each bead—the spring force, the frictional force due the systematic motion of the medium relative to the bead, and the osmotic force or driving force for diffusion which arises from the random thermal agitation of the system — provides an expression for $\dot{r}$ in Eq.(4.1). The result is a $3(N+1)$ dimensional diffusion equation for Ψ (100).,

The stress contributed by the chains is given by[1]:

$$p_{jk} = v \sum_{m=0}^{N} \langle F_j x_k \rangle_m \tag{4.2}$$

in which F_j is the j component of frictional force exerted on the medium by the bead, x_k is the k component of the position vector of the bead, and $\langle\ \rangle_m$ means the average for bead m over the distribution function Ψ. Calculation of these averages requires a transformation to normal coordinates to separate the diffusion equation into a sum of terms, each depending on only one of the coordinates. Each term in the new coordinate system governs the response of one of $3(N+1)$ normal modes of the chain. Three modes are external, describing the diffusive and convective displacements of the entire chain; the remaining $3N$ (N for each coordinate direction) are internal modes, describing the relative motions of the beads.

The required transformation is that which diagonalizes the $N+1 \times N+1$ matrix:

$$A = \begin{bmatrix} 1 & -1 & 0 & 0 & \cdots\cdots\cdots\cdots\cdots & 0 \\ -1 & 2 & -1 & 0 & \cdots\cdots\cdots\cdots \\ 0 & -1 & 2 & -1 & \cdots\cdots\cdots\cdots \\ & & \cdots\cdots & 0 & -1 & 2 & -1 & & 0 \\ & & \cdots\cdots & 0 & 0 & -1 & 2 & & -1 \\ & 0\cdots\cdots\cdots & & 0 & 0 & 0 & -1 & & 1 \end{bmatrix} \cdot \tag{4.3}$$

For N large and $i \ll N$ the eigenvalues of A are

$$\lambda_i = \left(\frac{\pi i}{N}\right)^2 \qquad i = 0, 1, 2, \ldots . \tag{4.4}$$

---

[1] This equation has recently been derived in rather complete fashion for bead-spring models by Lodge and Wu (101). Derivation and application to models other than spring-bead systems is given by Bird and co-workers (102).

The associated relaxation times of the internal modes are

$$\tau_i = \frac{\langle r^2 \rangle n \zeta_0}{6\pi^2 kT} \frac{1}{i^2} \qquad i = 1, 2, \ldots . \tag{4.5}$$

The internal modes are periodic functions of bead position along the chain; the characteristic wavelength of internal mode i is N/i. Thus the lower modes govern the large scale motions and have the longest relaxation times.

4.3.1. Configurational Relaxation

Configurational relaxation in the absence of flow is governed by the normal modes and their corresponding relaxation times. The autocorrelation function for the end-to-end vector is a measure of the configurational memory:

$$C(t) = \frac{\langle r(0) \cdot r(t) \rangle}{\langle r(0) \cdot r(0) \rangle} \tag{4.6}$$

in which $r(0) = r_N - r_0$ is the end-to-end vector at some initial time, and $r(t)$ is the vector for the same chain at a later time t. For the spring-bead model one can easily show:

$$C(t) = \frac{8}{\pi^2} \sum_{i=1, 3, 5, \ldots} \frac{1}{i^2} e^{-t/2\tau_i}. \tag{4.7}$$

The correlation soon becomes dominated by the relaxation of the lowest mode:

$$C(t) = \frac{8}{\pi^2} e^{-t/2\tau_1}. \tag{4.8}$$

The relaxation of $\langle r^2 \rangle$ towards its equilibrium value $\langle r^2 \rangle_e$ for chains with an initial value $\langle r^2 \rangle_0$ is given by an equation similar to Eq.(4.7), and is similarly dominated by τ_1 (103). Furthermore, for chains which are all stretched initially to the same end separation r_0 and released, the relaxation of $\langle r^2 \rangle$ to the equilibrium value $\langle r^2 \rangle_e$ is given by (103):

$$\langle r^2 \rangle = r_0^2 C^2(t) + \langle r^2 \rangle_e [1 - C(2t)] \tag{4.9}$$

where $C(t)$ is given by Eq.(4.7). Again, like the autocorrelation function, the recovery to $\langle r^2 \rangle_e$ is soon dominated by τ_1. In general, one might expect any measure of the configurational memory time of the chains to be of the order of the relaxation time of the lowest normal mode.

4.3.2. Response to Flow

The configurational response to flow depends upon which of the normal modes interact frictionally with the flow field. In simple shear the distribution envelope in the flow direction alone is altered, and only the N normal modes associated with the flow direction are active. The polymer contribution to the shear relaxation modulus for a system with v chains per unit volume is:

$$G(t) = vkT \sum_{i=1}^{N} e^{-t/\tau_i} \tag{4.10}$$

in which τ_i is the relaxation time associated with the ith internal mode. The steady shear viscosity and recoverable shear compliance in dilute solutions are therefore:

$$\eta - \eta_s = vkT \sum_{i=1}^{N} \tau_i, \tag{4.11}$$

$$J_e^0 = \frac{1}{vkT} \frac{(\eta - \eta_s)^2}{\eta^2} \frac{\sum_{i=1}^{N} \tau_i^2}{\left(\sum_{i=1}^{N} \tau_i\right)^2} \tag{4.12}$$

where η_s is the solvent viscosity. These expressions can be converted to macroscopic variables by replacing vkT by cRT/M.

The validity of Eqs.(4.10)–(4.12) probably extends well beyond the Rouse model itself [characterized by the specific set of τ_i values in Eq.(4.5)], and it seems likely that they will apply, at least for small disturbances, whenever the elements supporting the stress are joined by sufficiently flexible connectors and configurational relaxation is driven by simple Brownian diffusion. One might speculate further that these same forms would apply even in concentrated systems, with Eq.(4.10) expressed in a somewhat more general form because of intermolecular interactions:

$$G(t) = \frac{1}{3} kT \sum_{i=1}^{N_T} e^{-t/\tau_i}. \tag{4.13}$$

In Eq.(4.13) N_T is the total number of internal degrees of freedom per unit volume which relax by simple diffusion ($N_T = 3\nu N$ for dilute solutions), and τ_i is the relaxation time of the ith normal mode ($i = 1, 2, ..., 3N_T$) for small disturbances. Equation (4.13), together with a stipulation that all relaxation times have the same temperature coefficient, provides, in fact, the molecular basis of time-temperature superposition in linear viscoelasticity. It also reduces to the expression for the equilibrium shear modulus in the kinetic theory of rubber elasticity when $\tau_i = \infty$ for some of the modes.

Zimm (100) has extended the Rouse model to allow for intramolecular hydro-dynamic interaction, i.e., changes in medium velocity near each bead caused by the flow disturbance from other beads on the same chain. The Oseen approxima-tion, evaluated with the beads located at their mean equilibrium positions, was used to estimate the velocity disturbances. The intensity of the disturbance depends on the parameter h:

$$h = \frac{n\zeta_0}{(12\pi^3)^{1/2}\eta_s\langle r^2\rangle^{1/2}}.$$ (4.14)

Note that h is proportional to $n^{1/2}$ in θ-solvents, and thus to $N^{1/2}$. For $h = 0$ the flow disturbance is zero, the chain is said to be free draining, and the original Rouse model is recovered. For $h \gg 1$, flow in the coil interior is presumed to be substantially reduced, the chain is frequently said to behave as an impenetrable coil, and the Zimm model is obtained. Equations (4.10–4.12) continue to apply for all values of h, although the distribution of relaxation times depends on h.

Some results for the two limiting cases and large N are:

$h = 0$ (Rouse model)

$$\tau_i = \frac{6}{\pi^2}\frac{\eta_0 - \eta_s}{\nu kT}\frac{1}{i^2},$$ (4.15)

$$\eta_0 - \eta_s = \frac{\zeta_0\langle r^2\rangle vn}{36},$$ (4.16)

$$J_e^0 = \frac{2}{5}\frac{1}{\nu kT}\left(\frac{\eta_0 - \eta_s}{\eta_0}\right)^2.$$ (4.17)

$h \gg 1$ (Zimm model)

$$\tau_i = \frac{6}{\pi^2}\frac{\eta_0 - \eta_s}{\nu kT}\frac{1}{b_i}, \qquad b_i = 1.44, 4.55, 8.60, 13.5, ...,$$ (4.18)

$$\eta_0 - \eta_s = \left(\frac{0.877}{h}\right)\left(\frac{\zeta_0\langle r^2\rangle vn}{36}\right) = 2.84 \times 10^{23}\frac{\langle r^2\rangle^{3/2}}{M}\eta_s c,$$ (4.19)

$$J_e^0 = 0.206\frac{1}{\nu kT}\left(\frac{\eta_0 - \eta_s}{\eta_0}\right).$$ (4.20)

The Zimm model is most appropriate for long flexible chains in dilute solution. Tschoegl (104) has calculated eigenvalues for intermediate values of h and non-theta solvents. Fixman and Pyun (105) used a perturbation scheme to allow the elements in the Oseen matrix to change as the coil deforms, obtaining thereby slightly different eigenvalues for $h \gg 1$. Thurston and Morrison (106) have evaluated the pre-averaged eigenvalues for unperturbed dimensions more precisely and have examined the effect of the number of subchains on the results. Lodge and Wu (107) recently extended the calculations to $N = 300$ for a range of values of $h^* = h/N^{1/2}$.

Constitutive equations for the Rouse and Zimm models have been derived, and are found to be expressible in the form of Lodge's elastic liquid equation [Eq.(6.15)], with memory function given by (101):

$$M(t - t') = vkT \sum_{i=1}^{N} \frac{e^{-(t-t')/\tau_i}}{\tau_i}. \qquad (4.21)$$

In steady shear flow, the viscosity is independent of shear rate: $\eta = \eta_0$ for all $\dot{\gamma}$. This property alone represents a serious qualitative failure of the conventional bead-spring models. The normal stress functions are (108):

$$p_{11} - p_{22} = N_1 = \frac{4}{5} \frac{M}{cRT} (\eta_0 - \eta_s)^2 \dot{\gamma}^2 \quad (h = 0), \qquad (4.22)$$

$$p_{11} - p_{22} = N_1 = 0.410 \frac{M}{cRT} (\eta_0 - \eta_s)^2 \dot{\gamma}^2 \quad (h = \infty), \qquad (4.23)$$

$$p_{22} - p_{33} = N_2 = 0 \quad \text{(all values of } h). \qquad (4.24)$$

The Rouse form ($h = 0$) is frequently applied to situations in which the solvent contribution to the viscosity is negligible. For this case,

$$\eta_0 = \left[\frac{\zeta_0 \, c(\langle r^2 \rangle / M)}{36 N_a \, m_0} \right] M, \qquad (4.25)$$

$$J_e^0 = \frac{2}{5} \frac{M}{cRT} \qquad (4.26)$$

in which m_0 is the molecular weight per main chain atom and N_a is Avogadro's number. In polydisperse systems (109):

$$\eta_0 = \left[\frac{\zeta_0 \, c(\langle r^2 \rangle / M)}{36 N_a \, m_0} \right] \bar{M}_w, \qquad (4.27)$$

$$J_e^0 = \frac{2}{5} \frac{\bar{M}_w}{cRT} \frac{\bar{M}_z \bar{M}_{z+1}}{\bar{M}_w^2}. \qquad (4.28)$$

The various average molecular weights are defined as ratios of successive moments of the weight distribution function $W(M)$: $Q_k = \int\limits_0^\infty M^k W(M)\, dM$ and $\bar{M}_w = Q_1/Q_0$, $\bar{M}_z = Q_2/Q_1$, and $\bar{M}_{z+1} = Q_3/Q_2$. The steady state values of η_0 and J_e^0 are independent of the number of elements in the bead-spring model as long as N is sufficiently large. However, the initial modulus depends explicitly on N:

$$G^0 = \nu N k T = \frac{N c R T}{M} \tag{4.29}$$

confirming the expectation that short time response is sensitive to structural details of the model, and thus making spring-bead models inappropriate for describing the very rapid relaxation processes.

The relaxation time distribution of the bead-spring models is discrete. The spectrum is

$$H(\tau) = \nu k T \sum_{i=1}^N \delta(\tau - \tau_i) \tag{4.30}$$

in which $\delta(\ )$ is the Dirac delta function. For the Rouse model with negligible solvent contribution and N large:

$$\tau_i = \tau_1/i^2, \tag{4.31}$$

$$\tau_1 = \frac{6}{\pi^2} \frac{\eta_0 M}{c R T}. \tag{4.32}$$

The form of the spectrum is the same for Rouse chains fixed at one or both ends, although τ_1 in these cases is no longer given by Eq. (4.32). This is also the spectrum at long times for various models with explicit forms for the local motions (91, 92). The Rouse spectrum can also be represented by a continuous function for times which are small compared to τ_1. For $N \to \infty$:

$$H(\tau) = \frac{\nu k T}{2} \left(\frac{\tau_1}{\tau}\right)^{1/2} \qquad \tau < \tau_1/a^2$$

$$H(\tau) = 0 \qquad\qquad \tau > \tau_1/a^2. \tag{4.33}$$

The cut off parameter a is of order unity. Its value is somewhat arbitrary, reflecting the inability of this continuous spectrum to represent the long time behavior of the Rouse model precisely. Thus, with Eq. (4.33) and Eqs. (3.24) and (3.25):

$$\eta_0 = \frac{1}{a} \frac{c R T}{M} \tau_1, \tag{4.34}$$

$$J_e^0 = \frac{1}{3a} \frac{M}{c R T}, \tag{4.35}$$

and no value of a will give the correct numerical coefficients in both equations ($\pi^2/6$ and $2/5$, respectively).

4.4. Properties of Alternative Models

Other models have also been used to represent polymer chains. The elastic dumbbell concentrates all frictional resistance in two beads which are joined by a linear spring, and is merely the Rouse model with $N = 1$. The spring constant K is chosen to provide the same equilibrium radius of gyration as a polymer molecule. The elastic dumbbell has many of the properties of the Rouse model, except that it provides only a single relaxation time. Computational simplicity makes it convenient for rough estimations. The rigid dumbbell also has two frictional sites, but they are separated by a connector of fixed length which may by chosen to match the radius of gyration of the polymer molecule. Unlike the linear spring models, the viscosity of the rigid dumbbell depends on shear rate. Reference (102) has a rather complete discussion of such models. Another model which one might consider is a rigid spherical cloud of frictional sites with the same mean spatial distribution of resistances as the undisturbed polymer molecule.

Curtiss et al. (102a) have recently developed general formulas for $\eta_0 - \eta_s$ and J_e^0 for free-draining bead-connector models with arbitrary numbers of beads, connecting arrangements and force-distance laws for the connectors. The expressions depend on averages over the unperturbed distribution of configurations for the model:

$$\eta_0 - \eta_s = \frac{\zeta \, n \, v \langle \xi_1 + \xi_2 + \xi_3 \rangle}{6}, \tag{4.36}$$

$$J_e^0 = \frac{3}{5} \left[\frac{\langle \xi_1^2 \rangle + \langle \xi_2^2 \rangle + \langle \xi_3^2 \rangle - \langle \xi_1 \xi_2 \rangle - \langle \xi_1 \xi_3 \rangle - \langle \xi_2 \xi_3 \rangle}{\langle \xi_1 + \xi_2 + \xi_3 \rangle^2} \right] \frac{(\eta_0 - \eta_s)^2}{v k T \, \eta_0^2}. \tag{4.37}$$

in which

$$\xi_k = \frac{1}{n} \sum_{i=1}^{n} (x_k)_i^2 \quad k = 1, 2, 3. \tag{4.38}$$

The number of beads in the model macromolecule is n, and ζ is the Stokes' law friction coefficient of each bead. The ξ_k are to be evaluated for each macromolecule in its own internal coordinate system, with origin at the molecular center of gravity and axes ($k = 1, 2, 3$) lying along the principal axes of the macromolecule. The coordinates of the ith bead in this frame of reference are $(x_1)_i$, $(x_2)_i$, and $(x_3)_i$. The averaging indicated by $\langle \; \rangle$ is performed over all macromolecules in the system. Thus, $\langle \xi_1 + \xi_2 + \xi_3 \rangle$ is simply S^2 for the macromolecules. The viscosity is therefore identical, for all free-draining models with the same molecular frictional coefficient $n\zeta$ and the same radius of gyration, to the expression from the Rouse theory:

$$\eta_0 - \eta_s = \frac{\zeta \, n \, v \, S^2}{6}. \tag{4.39}$$

In rigid models each macromolecule has the same set of ξ_k, so $\langle \xi_k \rangle = \xi_k$ and $\langle \xi_j \xi_k \rangle = \xi_j \xi_k$. For flexible models the averages are calculated from the equilibrium distribution of internal configurations (each configuration weighted according to its energy through the Boltzmann equation). Thus, one sees immediately that $J_e^0 \equiv 0$ for rigid models with sufficient symmetry to yield $\xi_1 = \xi_2 = \xi_3$, and hence for the rigid-spherical cloud model. Likewise in any model with only two beads, $\xi_2 = \xi_3 = 0$, and

$$ J_e^0 = \frac{3}{5} \frac{\langle \xi_1^2 \rangle}{\langle \xi_1 \rangle^2} \frac{(\eta_0 - \eta_s)^2}{vkT\,\eta_0^2}. \tag{4.40} $$

For the rigid dumbbell, therefore, $J_e^0 = \frac{3}{5}(\eta_0 - \eta_s)^2/vkT\,\eta_0^2$, while for the elastic (linear spring connector) dumbbell it is easy to show that $\langle \xi_1^2 \rangle / \langle \xi_1 \rangle^2 = 5/3$, so $J_e^0 = (\eta_0 - \eta_s)^2/vkT\,\eta_0^2$. It is clear that all rigid models with the same set of ξ_k and the same value of ζn give the same values of $\eta_0 - \eta_s$ and J_e^0, and that it is molecular asymmetry in rigid models (differences among ξ_1, ξ_2 and ξ_3) which governs the value of J_e^0. The numerical factor can never exceed $3/5$ in rigid models. Molecular flexibility contributes to J_e^0 also. Thus, from the nature of the averages in Eq. (4.37), one would expect J_e^0 always to be greater for a flexible model than for a rigid model whose ξ_k are taken equal to $\langle \xi_k \rangle$ ($k = 1, 2, 3$) of the flexible model.

These results make it clear that the forms of $\eta_0 - \eta_s$ and J_e^0 are completely independent of model details. Only the numerical coefficient of J_e^0 contains information on the properties of the model, and even then the result depends on both molecular asymmetry and flexibility. Furthermore, polydispersity effects are the same in all such free-draining models. The forms from the Rouse theory carry over directly, so that $\eta_0 - \eta_s$, translated to macroscopic terms, is proportional to $\bar{M}_w$ and J_e^0 is proportional to the factor $\bar{M}_z \bar{M}_{z+1}/\bar{M}_w$. Unfortunately, no such general analysis has been made for models with intramolecular hydrodynamic interaction, and of course these results apply in principle only to cases where intermolecular interactions are negligible.

Completely rigid models appear to provide rather peculiar short time response. The stress relaxation modulus for rigid dumbbells is (102):

$$ G(t) = vkT \left[\tfrac{2}{5}\,\delta(t) + \tfrac{3}{5}\,e^{-t/\tau} \right] \tag{4.41} $$

while that for elastic dumbbells is

$$ G(t) = vkT\,e^{-t/\tau}. \tag{4.42} $$

In each the relaxation time τ is $(\eta_0 - \eta_s)/vkT$, and each satisfies the requirement from linear viscoelasticity that

$$ \int_0^\infty G(t)\,dt = \eta_0 - \eta_s. \tag{4.43} $$

However, in performing this integration on experimental data it is found that behavior at short times tends to be relatively unimportant, so that

$$\int_{\varepsilon}^{\infty} G(t)\, dt \approx \eta_0 - \eta_s \tag{4.44}$$

over some range of sufficiently small ε. Elastic dumbbells behave this way for $\varepsilon \ll \tau$, since

$$\int_{\varepsilon}^{\infty} G(t)\, dt = (\eta_0 - \eta_s)\, e^{-\varepsilon/\tau} \tag{4.45}$$

but rigid dumbbells give

$$\int_{\varepsilon}^{\infty} G(t)\, dt = \tfrac{3}{5}(\eta_0 - \eta_s)\, e^{-\varepsilon/\tau} \tag{4.46}$$

and acquire the extra $\tfrac{2}{5}(\eta_0 - \eta_s)$ only when ε becomes literally zero. Thus, the experimental stress relaxation behavior of a solution of rigid dumbbells would appear to show large deviations from the laws of linear viscoelasticity. Similar anomalies would appear in the dynamic loss modulus.

The delta function anomaly is probably present in the stress relaxation modulus of all rigid models. It arises because the frictional sites in rigid models must move relative to the solvent in any small but arbitrarily rapid deformation, thereby generating a frictional force proportional to the deformation rate. In models with flexibility the frictional sites can move instantaneously with the fluid. The resisting force generated by the macromolecule is then simply that associated with connector stretches, the latter remaining bounded even for arbitrarily rapid deformation rates as long as the total deformation remained small. It seems clear therefore that models must contain some degree of flexibility to simulate the observed behavior of polymer systems.

4.5. Discussion

All calculations of visoelastic properties described here apply in principle only to dilute solutions, since no allowance for intermolecular interactions has been made. Nevertheless, the Rouse model in particular has been widely applied to concentrated systems. There is probably no fundamental justification for such an application. One simply assumes that each chain responds independently to the systematic motions of a medium which is composed of other chains and solvent, and which is taken to be a homogeneous Newtonian liquid (109). The contribution of the chains to the stress are taken to be additive.

The viewpoint parallels that of many other theories of condensed state behavior. The van der Waals theory develops an equation of state for dense gases from the assumption that each molecule moves in an average field provided by its neighbors and that the molecules contribute additively to the pressure. The Flory-Huggins thermodynamic theory of concentrated polymer solutions proceeds similarly. Chains select configurations on a lattice partially occupied by other chains and contribute additively to the configurational entropy. The cell theory for flow in concentrated suspensions is developed along a similar line. Each such theory seeks to capture the essential features of concentrated systems, although without necessarily expecting agreement in detail. The Rouse theory is qualitatively correct in many respects in concentrated systems. It therefore provides a convenient framework for the empirical organization of experimental data. It is also an appropriate starting point for theories which attempt to explain linear viscoelasticity on a more quantitative molecular basis and to remedy its failure to predict such non-linear properties as the shear-rate dependence of viscosity.

Although both the Rouse and Zimm models provide for chain connectivity in the sense that motions of individual beads are felt by adjacent beads, these models do not rule out motions in which one section of chain literally crosses through another. The success of the Zimm model in dilute solutions implies that restrictions on the motion due to connectivity, *i.e.*, self entanglement, are relatively unimportant in dilute solutions. The reason for this may be that hydrodynamic interaction already tends to rule out relaxation paths through the coil interior, and it is just these paths for which self-entanglement effects would be most prevalent. In Monte Carlo studies of lattice chains, Verdier (*110*) noted a considerable slowing in configurational relaxation when the chain was programmed for excluded volume (non-simultaneous occupancy of lattice sites). In this case volume exclusion disallows chain crossing as well, and so perhaps allows the display of an effect caused by self-entanglement, unobscured by hydrodynamic interaction but unobservable in systems of real chains in dilute solution.

5. Experimental Results on Linear Viscoelastic Behavior

5.1. Introduction

The effects attributed to entangling interactions, *e.g.*, the plateau region in stress relaxation, appear most prominently at high concentrations and in melts. It is important, however, to distinguish this interaction from other types which are present at lower polymer concentrations. To make the separation properly, it is necessary to examine viscoelastic behavior at all levels of concentration, beginning at infinite dilution.

5.2. Very Dilute Solutions

The importance of intramolecular hydrodynamic interaction at low polymer concentrations is well established. The impenetrable sphere form (*20, 27*):

$$[\eta] = \lim_{c \to 0} \frac{\eta_0 - \eta_s}{\eta_s c} = \Phi_\infty \frac{\langle r^2 \rangle^{3/2}}{M} \tag{5.1}$$

relates intrinsic viscosity, molecular dimensions and molecular weight at high molecular weights $(M > 10^6)$ for a wide variety of polymer-solvent systems (19, 111). In theta solvents the experimental values of $[\eta] M/\langle r^2 \rangle^{3/2}$ remain very nearly constant even down to rather low molecular weights $(M \sim 10^4)$, while, broadly speaking, in good solvents they already begin to decrease as the molecular weight falls below the range $M \sim 10^5$. The observed value of Φ_∞ agrees well with theoretical limiting values $(h \to \infty)$ from the original Zimm theory, $\Phi_\infty = 2.84 \times 10^{23}$, or the Pyun-Fixman modification, $\Phi_\infty = 2.68 \times 10^{23}$. The decrease of Φ with decreasing molecular weight is interpreted as a trend toward free-draining behavior, and is expected theoretically since h [Eq. (4.14)] itself decreases with decreasing molecular weight, being proportional to $M/\langle r^2 \rangle^{1/2}$.

Information on dynamic behavior at infinite dilution has greatly increased in recent years through the development of instruments for precise determination of $G'(\omega)$ and $G''(\omega)$ in dilute solutions (112, 113). Ferry and coworkers have published extensively on the intrinsic dynamic moduli,

$$[G']_\omega = \lim_{c \to 0} \frac{G'(\omega)}{c},$$
(5.2)

$$[G'']_\omega = \lim_{c \to 0} \frac{G''(\omega) - \eta_s \omega}{c}.$$
(5.3)

Data on several polymers in both good solvents and theta solvents are now available (114–116). With h^* as an adjustable parameter, reduced plots of $[G']_\omega M/RT$ and $[G'']_\omega M/RT$ vs $[\eta]\eta_s\omega/RT$ can be placed in essentially quantitative agreement with moduli calculated from the Zimm eigenvalues (107, $116a$) for large numbers of submolecules (Fig. 5.1). Highly accurate calculations have now been carried out for up to $N = 200$ submolecules. The deduced values of h_{200}^* vary somewhat with solvent power, but results for different polymers, molecular weights and solvent powers are brought together by the correlation (116):

$$h_{200}^* \alpha_\eta = 0.21 \pm 0.02$$
(5.4)

in which α_η is the viscometric coil expansion factor $[[\eta]/[\eta]_\theta]^{1/3}$. This form is suggested by the $\langle r^2 \rangle^{1/2}$ term in the denominator of Eq. (4.14) in combination with Eq. (5.1). Thus the inference is drawn that coil expansion, brought about by excluded volume in good solvents, shifts the dynamic properties toward free draining behavior, and that the net result can be understood simply in terms of a uniformly expanded gaussian chain and a proportionate reduction of h.

Systematic departures from the Zimm moduli are observed at high frequencies (93, 117). These deviations appear to stem from the expected inadequacies of spring-bead models when the driving frequency approaches the frequency of the primitive backbone motions. The effects are attributed to a local resistance to the articulations of the chain which are required to bring about configurational

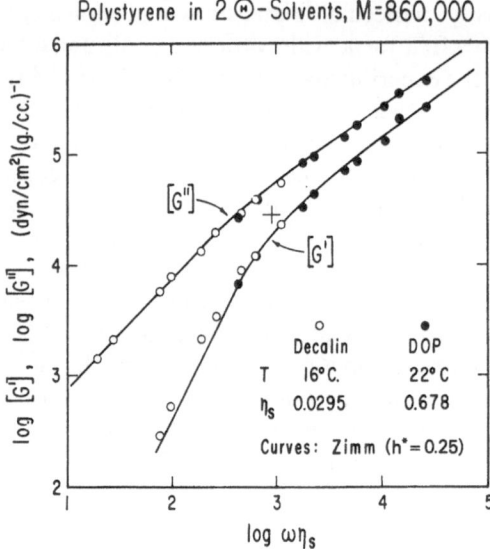

Fig. 5.1. Intrinsic moduli for narrow distribution polystyrene ($M = 860000$) in two theta solvents (*114*). This comparison with theory is equivalent to that of reduced moduli described in the text. [Reproduced from Polymer J. **1**, 747 (1970).]

change. They are well correlated by an internal viscosity or frictional coefficient operating on the normal modes (*118*). Experimental values for the internal viscosity coefficient are insensitive to molecular weight as expected, but appear to vary in direct proportion to the solvent viscosity. A review of these results has recently appeared in this journal (*118a*).

5.3. Concentration Effects and Intermolecular Interaction

Fundamental theories of transport properties for systems of finite concentration are still rather tentative (*24*). The difficulties are accentuated by the still uncertain effects of concentration on equilibrium properties such as coil dimensions and the distribution of molecular centers. Such problems are by no means limited to polymer solutions however. Even for the supposedly simpler case of hard sphere suspensions the theories of concentration dependence for the viscosity are far from settled (*119, 120*).

The Huggins constant k' characterizes the first effects of interaction on the zero-shear viscosity:

$$\eta_0 = \eta_s(1 + [\eta]\,c + k'\,[\eta]^2\,c^2 + \cdots).\tag{5.5}$$

Experimentally k' is essentially independent of molecular weight for long chains, with values of roughly 0.30–0.40 in good solvents and 0.50–0.80 in theta solvents.

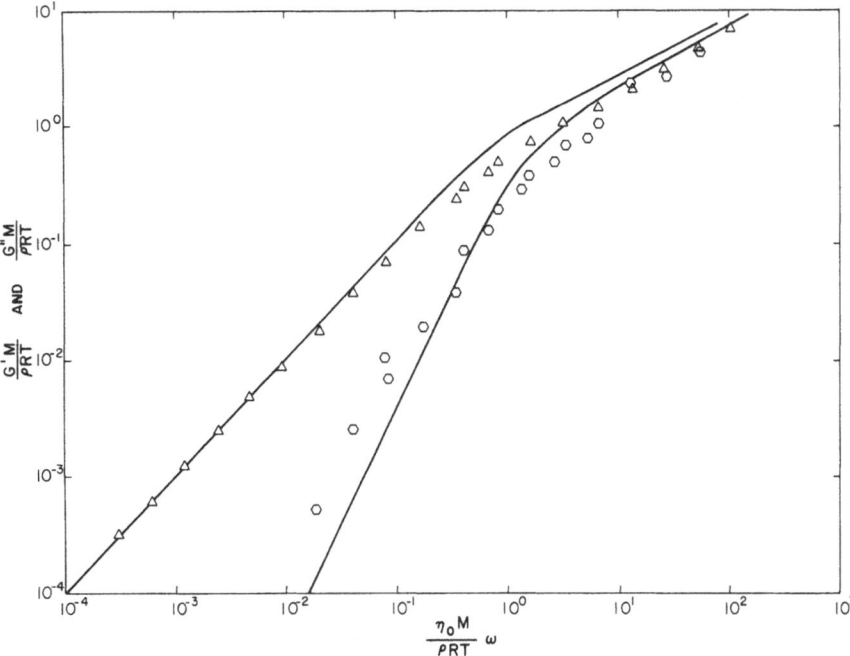

Fig. 5.2. Reduced dynamic moduli for undiluted narrow distribution polystyrene of low molecular weight. Data for a sample of $\bar{M}_w = 28900$ were reduced to 160 °C, for which $\varrho = 1.0$ gm/ml and $\eta_0 = 54500$ poise (124). The solid lines were calculated from the Rouse theory

Theories of k' for random coil molecules are very difficult and still somewhat lacking in experimental confirmation (24, 121).

Fundamental theories of the relaxation spectrum at finite concentration have yet to be developed. Experimentally, intermolecular contributions appear quickly as the concentration increases. In the dynamic moduli they appear first as an increase in the longest relaxation time, the more rapid relaxations remaining relatively unaffected (114). Apparently the largest scale molecular motions are slowed as peripheral segments begin to impinge upon those of neighboring chains. At somewhat higher concentrations, as the coils begin to overlap appreciably, the faster relaxations also shift successively to longer times. The relative spacings at long times increasingly resemble the Rouse spectrum (15, 122, 123). This behavior is believed to reflect an increasing free-draining character in the local flow patterns, caused by a gradual cancellation of intramolecular hydrodynamic effects as the solution becomes uniformly filled with polymer segments.

The culmination of this trend is illustrated in Fig. 5.2 by dynamic data on undiluted polystyrene of low molecular weight (124). Agreement with the Rouse model here is by no means as good as that seen in Fig. 5.1 with the Zimm model for a high molecular weight polystyrene at infinite dilution. Indeed, the value of J_e^0 deduced from $G'(\omega)$ for the sample in Fig. 5.2 exceeds the value from the Rouse theory by approximately a factor of two. Nevertheless, the agreement is remarkable, considering that no arbitrary parameters are involved (aside from the fact that the reduction scheme compels agreement with the observed value of η_0) and that intermolecular interactions have not been considered at all. Moreover, even the magnitude of η_0 in this case is probably near that given by the unmodified Rouse theory (see Part 5.4.1). Thus, although intermolecular interaction can be important even in dilute solutions, this importance does not necessarily carry over into concentrated solutions and the melt state, at least for short chains.

5.3.1. Parameters Measuring Intermolecular Interactions

The literature on viscosity at low and high concentrations of polymer suggests very clearly that two fundamentally different types of intermolecular interaction need to be considered. The viscosity behavior at low concentrations resembles in many ways that of a suspension of discrete rigid particles (119, 120). The molecules probably retain some vestige of non-draining character in this regime, and intermolecular effects may be caused by the interaction between the flow fields of the molecules, as in the case of rigid particles. The interaction seems to depend upon the volume occupied by the molecules or, alternatively, on the degree of overlap of the individual molecular domains. Frisch and Simha (125) have pointed out that the parameter $c[\eta]$ measures the degree of coil overlap in a solution, if one neglects changes in coil dimensions with concentration. The pervaded volume V of a molecule with radius of gyration S may be estimated roughly as $4\pi S^3/3$. In combination with Eq. (5.1) and with $S^2 = \langle r^2 \rangle/6$ for random coils,

$$V = \frac{4\pi [\eta] M}{3 \Phi_\infty 6^{3/2}} .$$

(5.6)

The number of molecules per unit volume v is $6.02 \times 10^{23} c/M$. If the molecular centers are distributed randomly, the product vV is the average number of other molecules with centers lying within the pervaded volume of any one molecule. Accordingly, vV is a measure of the potential degree of coil overlap, and with $\Phi_\infty = 2.68 \times 10^{23}$:

$$vV = 0.64 \, c[\eta] .$$

(5.7)

Simha and Zakin (126), Onogi et al. (127), and Cornet (128) develop overlap criteria of the same form but with different numerical coefficients. Accordingly, flow properties which depend on concentration and molecular weight principally through their effects on coil overlap should correlate through the Simha parameter $c[\eta]$, or cM^a, in which a is the Mark-Houwink viscosity exponent ($0.5 < a < 0.8$). If coil shrinkage, caused by the loss of excluded volume in good solvents, is itself a function of coil overlap, then $c[\eta]$ would remain an acceptable correlating parameter even in good solvents.

A second source of intermolecular interactions, segment-segment contacts between molecules, has been considered in concentrated solutions and melts where free draining flow is assumed to dominate. This appears to be the mode of interaction envisioned by Bueche in his discussion of chain entanglement (7). In a system of uniform segment density the number of intermolecular contacts per unit volume is proportional to c^2. Since the number of polymer molecules per unit volume is proportional to c/M, the number of intermolecular contacts per molecule is proportional to cM. Thus, properties which depend on c and M principally through their effects on the number of segment-segment contacts per molecule should correlate in terms of the Bueche parameter cM. In free draining systems the variation of ζ_0 with polymer concentration must be accounted for separately of course. Since $c[\eta]$ and cM involve different combinations of c and M, it should in principle be possible to distinguish between Simha and Bueche interactions.

5.3.2. Viscosity Correlations

The effects of interaction on viscoelastic properties at low concentrations depend on the Simha parameter. For example, Ferry has pointed out the importance of $c[\eta]$ for the transition from Zimm-like to Rouse-like behavior in the dynamic properties and in the observed values of J_e^0 (15). The shear rate dependence of viscosity undergoes a corresponding transition as a function of $c[\eta]$ (see Part 8).

The usefulness of $c[\eta]$ as an *approximate* reducing parameter for the zero shear viscosity at moderate concentrations is very well established. Master equations of the form

$$\eta_0 = \eta_s\, F(c[\eta]) \tag{5.8}$$

are fairly successful in unifying data for different concentrations and molecular weights in the same polymer-solvent system. One such expression is the Martin equation (129):

$$\eta_0 - \eta_s = \eta_s\, c[\eta]\, e^{k'\, c[\eta]}. \tag{5.9}$$

Somewhat improved reductions can be obtained by using $cM^{a'}$ as the independent variable, where a' is chosen empirically. In these cases a' is usually rather close to the Mark-Houwink exponent for the system (129–133).

Cornet (128) points out that plots of $\log \eta_0$ vs $\log c$ tend to be curved at low concentrations but become straight lines at high concentrations. The concentration at the onset of straight line behavior, designated c^*, was assumed by Cornet to be the concentration at which a uniform segment density is attained in the solution. He calculates theta dimensions for several polymers from c^*, finding reasonably good agreement with literature values. With Eq. (5.1), his relation between c^* and $\langle r^2 \rangle_\theta$ can be cast in the form:

$$c^* = 6.14/[\eta]_\theta . \tag{5.10}$$

According to Cornet's criterion, the product $c^* M^{1/2}$ should be a constant for a given polymer-solvent system.

Onogi and coworkers (132, 133) have emphasized the importance of $c[\eta]$ (or $cM^{a'}$) in separating regions of weak and strong concentration dependence in η_0. The transitions on plots of $\log \eta_0$ vs $\log cM^{a'}$ are gradual and the shapes of the plots vary considerably from one polymer system to another. Figure 5.3 shows the behavior of several polymer-solvent systems plotted as suggested by the Martin expression [Eq. (5.9)]. In this form of plot no obvious transition is observable. It is therefore difficult to assign a critical value to ($cM^{a'}$). Nevertheless, the development of strong interactions, as judged by the viscosity, is clearly controlled by the overlap parameter $c[\eta]$, or $cM^{a'}$ for a given polymer-solvent system.

Williams and Gandhi (121, 134) have recently examined the effects of solvent power and polymer polarity on viscosity. They find that η_0 tends to increase with concentration more rapidly in theta-solvents than in good solvents and attribute this result to increased polymer-polymer association in thermodynamically poor solvents. Dreval and co-workers (135) and Ferry and co-workers (136) had earlier noted similar differences between good and poor solvents and drew similar inferences. On the other hand, Simha and Zakin (137) have pointed out that the least part of the solvent power effect may be accounted for by the expected decrease of coil dimensions in good solvents with increasing concentration. Thus in good solvents the coils are larger at low concentration due to excluded volume, and the relative viscosity of their solutions η_0/η_s will be larger. At higher concentrations the coil dimensions in good solvents shrink toward their theta-solvent values. The relative viscosity might therefore be expected to become the same at the same polymer concentration in different solvents.

Quadrat and Podnecka (138) have shown in a recent examination of literature data that such ideas account very well for the behavior of polyisobutylene in a variety of solvents. The results of Ferry and co-workers on polyisobutylene also fall in line when relative viscosities are compared (139). However, in polyvinyl acetate (136) the relative viscosities for a theta solvent cross over and became somewhat higher than good solvent values at higher concentrations ($> 35\%$ polymer). For polystyrene (121, 137, 138, 140) theta solvents also give higher relative viscosities at high concentration, with crossover concentrations in the range 15–30% polymer. For polymethyl methacrylate (121, 140a) the solvent effect is dramatic, with a crossover concentration in the range of 5–7% polymer.

Some relationship between viscosity crossover in theta solvents and polymer polarity is suggested by the results, supporting the idea of enhanced intermolecular association in poor solvents. However, from the data on hand, one could also infer a correlation with the glass transition temperature T_g of undiluted polymer,

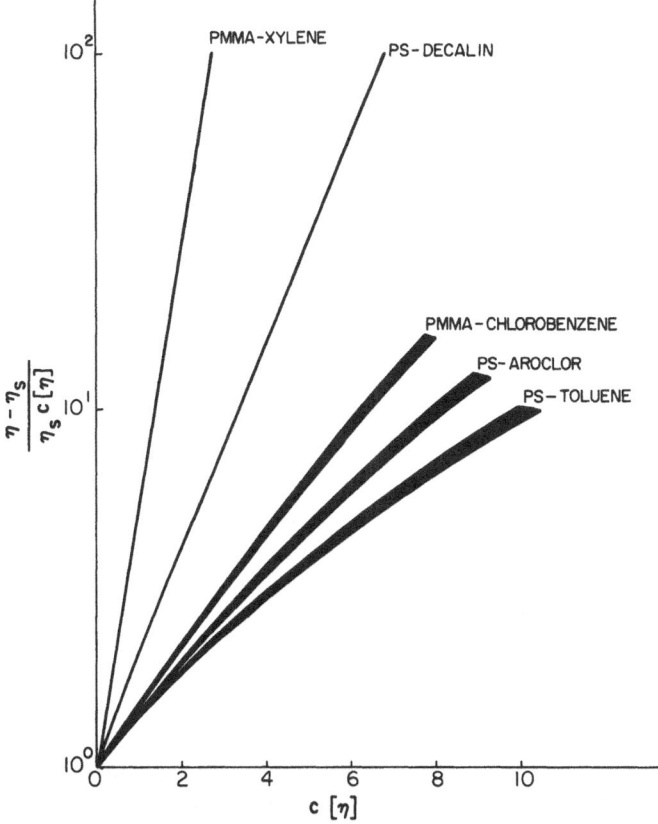

Fig. 5.3. Viscosity at various concentrations and molecular weights in the low to moderate concentration range. Polystyrene-decalin and polymethyl methacrylate-xylene are theta or near-theta systems; the remainder are good solvent systems (121, 177). Note that the $c[\eta]$ reduction is somewhat better in theta solvents, and that the Martin equation [Eq. (5.9)], which would give a straight line in the figure, is a somewhat better representation for theta solvents

and hence with T_g for the polymer-solvent mixture. Vinogradov and co-workers (141) have recently emphasized that such a dependence might develop as free draining behavior is approached at the higher concentrations, since the viscosity must lose its direct proportionality to solvent viscosity and become proportional instead to a local frictional coefficient ζ_0. The value of ζ_0 depends on the nature of both solvent and polymer since both influence the free volume and glass temperature of the mixture. The glass temperature of most solvents is well below room temperature. In polyisobutylene solutions T_g probably remains relatively constant and always well below room temperature because T_g for pure polyisobutylene is so low ($T_g = -70°$ C). For pure polyvinyl acetate T_g is $30°$ C, so T_g for its solutions must rise somewhat with increasing concentrations. The effect for polystyrene ($T_g = 95°$ C) should be larger, while that for polymethyl methacrylate

($T_g = 100°$ C) should not only be large but might be unusual if the unusual dependence of T_g with concentration in diethyl phthalate solutions (16, 142) holds true in other solvents as well.

Variations in the temperature coefficient of viscosity with solvent, which have also been presented as evidence of association in concentrated solutions (135, 143), could be similarly related to differences in T_g among the solutions. When free draining behavior is a possibility, the relative viscosities in different solvents should be compared at the same value of ζ_0 for the mixtures (that is, at constant free volume rather than at constant temperature). In any case, it is clear that a very well planned series of experiments is necessary in order to test for the existence of additional specific effects such as association. These comments are not meant to suggest that association can not occur at moderate concentrations. Indeed, the existence of association in various forms of polymethyl methacrylate seems well established (144). The purpose is rather to advocate that less specific causes be eliminated before association is inferred from viscosity measurements alone.

5.3.3. Relationships between Low and High Concentration Regimes

An important and unanswered question is just how the low and high concentration regimes fit together (130). For long chains the viscosity changes from dependence on $c[\eta]$ at moderate concentrations to dependence on the product of a local friction factor and a large-scale structure factor at high concentrations (16). In many of the studies already cited the reducing property of the parameter $c[\eta]$ persists to high concentrations ($> 30\%$ polymer), but this may be fortuitous. Free draining behavior combined with a mild concentration dependence in the friction factor could produce a similar correlation form [see Ref. (16), p. 303].

Perhaps the two flow regimes can be distinguished at least conceptually in the following way. When steady motions are imposed on the boundaries of a fluid, the flow pattern established in the interior is that which minimizes the rate of energy dissipation per unit volume $\dot{E}_v$ (145). In steady shearing flow $\dot{E}_v = \eta \dot{\gamma}^2$, so the local flow pattern in a polymer solution will be that which yields a minimum value for the viscosity. This principle then dictates the extent to which local lines of flow are diverted around the individual pervaded volumes. Non-draining or partial draining behavior is observed at infinite dilution because it results in a lower dissipation rate (lower viscosity) than free draining behavior.

Let us suppose that the experimental contribution of polymer to solution viscosity at low concentration, $\eta_0 - \eta_s$, can be represented to a sufficient approximation by the Martin equation [Eq. (5.9)]. An alternative which is presumably always available to the solution is free draining behavior [Eq. (4.16)]. As concentration is increased, then, according to the minimum dissipation principle, the behavior would be expected to change from the Martin form to the free draining form in the range where the two expressions for $\eta_0 - \eta_s$ become equal:

$$\frac{\zeta_0 \langle r^2 \rangle vn}{36} = \eta_s c[\eta] \, e^{k'c[\eta]}. \tag{5.11}$$

If $[\eta]$ in the front factor on the right side of Eq.(5.11) is replaced by its value according to the Zimm theory [Eq.(5.1)], and· if ζ_0 is replaced by the Stokes formula for spheres, $\zeta_0 = 6\pi\eta_s R_0$, in which R_0 is the effective Stokes radius of a main chain atom, Eq.(5.11) becomes:

$$1.1 \frac{nR_0}{\langle r^2 \rangle^{1/2}} = e^{k'c[\eta]}. \tag{5.12}$$

Thus, if R_0 is approximately equal to the length of a main chain bond, the left side of Eq.(5.12) is essentially the ratio of the contour length of a random coil to its mean end-to-end distance, a quantity which depends on solvent power and chain length, and which typically ranges from perhaps 10 to 100. Free draining then begins to dominate in a critical range of $c[\eta]$ values, given by Eq.(5.12):

$$(c[\eta])^* \approx \frac{1}{k'} \ln \frac{r_{extended}}{r_{coiled}} \approx 5\text{--}15. \tag{5.13}$$

These values provide the upper limit for a subrange of solution concentrations, beginning with the onset of coil overlap near $c[\eta] \approx 1$ and ending with essentially complete free draining character near $c[\eta] \approx 10$. This estimate of the upper limit is very approximate of course, since the behavior must also depend somewhat on solvent power and chain length. It is also subject to errors from the assumption that ζ_0 is essentially constant through the subrange, and that the free draining expression can be used without regard for other types of intermolecular interaction, such as chain entanglement.

5.3.4. Anomalous Flow Properties at Intermediate Concentrations

Time-dependent flow anomalies also appear in this region of moderate concentrations. Peterlin, for example, has observed that the viscosity of dilute solutions of various polymers may begin to increase in the course of flow through long capillaries (149). He attributes the effect to a shear-induced impingement and temporary aggregation of coils. Lodge (150) has observed macroscopic inhomogeneities in 2–3% solutions of polymethyl methacrylate and polystyrene after prolonged shearing in a plate and cone viscometer. Philippoff (151) has found that the flow birefringence of dilute polystyrene solutions is reduced by filtration, although the original values are restored after the solutions are allowed to stand for a time. Aggregation in dilute solutions which can be removed or reduced by heating is well established in certain highly polar polymers such as polyvinyl chloride. Not enough is known about such effects at the present time to warrant any detailed consideration here.

5.3.5. Summary

The viscoelastic properties of polymers in solutions of moderate concentration, roughly defined as the interval, $1 < c[\eta] < 10$, are not well understood in terms of fundamental theory. The difficulties parallel in some ways those of unifying the thermodynamic theories of very dilute and highly concentrated solutions. Broadly speaking, the molecules appear to retain some of their dilute solution impenetrable character. The choice of solvent influences viscosity both directly through its own viscosity and indirectly through its thermodynamic interactions with the polymer. Perhaps the most striking feature is the rather mild changes in relative spacings of the relaxation time distribution (the shift from a Zimm-like to Rouse-like spectrum) through the regime, even though the strength of the intermolecular interactions, as judged by viscosity and mean relaxation time, increases enormously.

It appears that one can develop a qualitative understanding of the simple flow properties at moderate concentration without going beyond concepts which are already inherent either in the dilute solution theory of polymers or in the properties of particulate suspensions. The dependence of viscosity on $c[\eta]$ is believed to reflect a particle-like or equivalent sphere (127) hydrodynamics in solutions of low to moderate concentration. In particular, the experimental facts do not force the consideration of effects which might arise from the permanent connectedness of the polymer backbones. Situations conducive to the entangling of molecules may be present, *e.g.*, overlap of the coils, but either entanglement contributions are small, or else they are overwhelmed by the $c[\eta]$ interactions.

5.4. Concentrated Solutions and Melts

Portions of the literature on viscoelasticity in concentrated polymer systems of narrow distribution have been reviewed recently (15, 16, 152, 153). The following discussion concerns three principal characteristics, the viscosity-molecular weight relation, the plateau modulus, and the steady-state compliance.

5.4.1. Zero-Shear Viscosity

A new set of flow characteristics gradually emerges as the concentration of polymer becomes large. The solution viscosity loses its direct dependence on solvent viscosity and comes to depend on the product of two parameters: a friction factor ζ which is controlled solely by local features such as the free volume (or alternatively the segmental jump frequency), and a structure factor F which is controlled by the large scale structure and configuration of the chains (16):

$$\eta_0 = \zeta F . \tag{5.14}$$

The Rouse expression for the viscosity [Eq. (4.25)] is of this form. The frictional coefficient per main chain atom ζ_0 is the friction factor in this case; the remainder of the expression is the structure factor.

The friction factor depends upon the same features that govern the viscosity of small-molecule liquids. At low temperatures ζ_0 depends on $T - T_g$ ($T_g < T < T_g + 100°$ C), and at higher temperatures it depends on an activation energy for flow. The value of T_g for a solution depends on the properties of both components and their concentrations, but it is independent of the large scale structure of the polymer as long as its molecular weight is large ($\bar{M}_n > 10^4$ for most linear polymers).

The structure factor, on the other hand, depends on the number of chains per unit volume and on their molecular weight and dimensions. Its behavior is largely the same from one polymer system to another, which is to say, independent of the chemical structure of the repeating unit. The Rouse equation, in fact, gives a good account of experimental results on linear chains if the molecular weight is not too large. Thus, after correction to constant friction factor, the viscosity for many undiluted polymers is nearly linear in molecular weight when the chain length is less than 300–500 main chain atoms (16). Moreover, the numerical values of ζ_0 deduced from η_0 with the Rouse equation and known values of $(S^2/M)_\theta$ are in quite reasonable accord with entirely independent estimates based on the diffusion coefficients of small molecules in the same system (15). Thus, for polymers of low molecular weight the Rouse model, with no adjustment for intermolecular interaction, gives not only the form but even the correct order of magnitude of the viscosity.[2]

For longer chains the form of the structure factor changes, becoming a much stronger function of molecular weight. At high molecular weights the well known 3.4 power dependence results:

$$\eta_0 = KM \qquad M < M_c$$
$$\eta_0 = K'M^{3.4} \qquad M > M_c. \tag{5.16}$$

The change in M dependence occurs smoothly but over a relatively narrow range in molecular weight. The characteristic value M_c is obtained from the intersection of straight lines drawn through the two branches of the curve. For many polymers M_c is independent of temperature and corresponds to a contour length of about 300–700 main chain atoms (16). In some systems the slope for $M > M_c$ appears to be somewhat higher than 3.4 (146). In others the slope begins to increase beyond 3.4 at higher molecular weights (147, 148) although in the case of polyvinyl acetate this may be caused by the presence of long chain branches. On the whole, however, the 3.4 power is supported by data from a remarkably wide range of linear polymers (16).

---

[2] It should be pointed out that this agreement cannot be used to support the Rouse model *per se*, since any model with the same total molecular friction and value of S^2 would give the same expression for η_0 (Part 4).

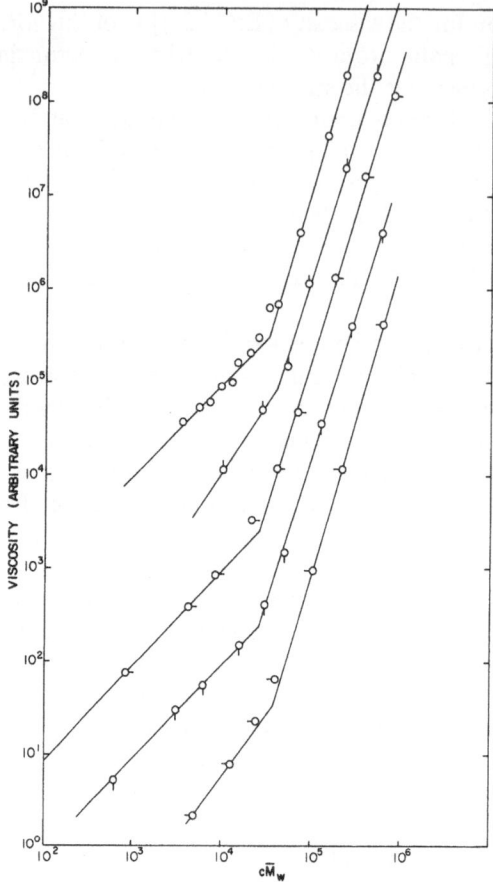

Fig. 5.4. Viscosity vs the product $c\bar{M}_w$ for polystyrenes at concentrations between 25% and 100%: ○, undiluted at 217 °C (154); ◐, 0.55 gm/ml in n-butyl benzene at 25 °C (155); ○–, 0.415 gm/ml in di-octyl phthalate at 30 °C (156); ○ (vertical), 0.310 gm/ml in di-octyl phthalate at 30 °C (156); and –○, 0.255 gm/ml in n-butyl benzene at 25 °C (155). Data at the various concentrations have been shifted vertically to avoid overlap

Concentrated solutions ($c >$ ca. 0.25 gm/ml) show similar behavior, although with characteristic molecular weights that increase with concentration. In many systems the transition remains fairly well defined and, within the accuracy of the data, conforms to

$$(M_c)_{\text{soln.}} = \left(\frac{\varrho}{c}\right) M_c = \frac{M_c}{\varphi} \tag{5.17}$$

where M_c is the value for undiluted polymer, ϱ is its density, and φ is the volume fraction of polymer in the solution. One can partially interpret this result by noting that the transition takes place for different concentrations at the same value of the segment contact parameter cM (see Part 5.3.1). Figure 5.4 shows η_0 vs cM for polystyrene melts (154) and polystyrene solutions at several con-

Table 5.1. Intersection points from Fig. 5.4 for η_0 vs cM in polystyrenes of various concentrations (visually estimated)

Concentration (gm/ml)	cM at intersection
1.00 (melt at 165° C)	32000
0.55 (n-butyl benzene at 30° C)	41000
0.415 (di-octyl phthalate at 25° C)	26000
0.310 (di-octyl phthalate at 25° C)	26000
0.255 (n-butyl benzene at 30° C)	38000

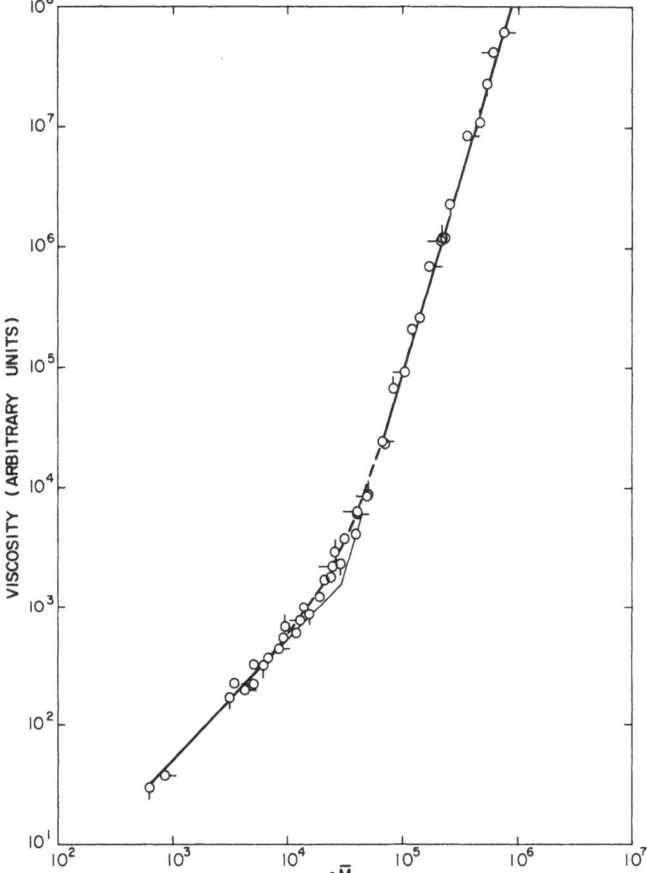

Fig. 5.5. Viscosity vs $c\bar{M}_w$ for polystyrenes at concentrations between 25% and 100%. The data in Fig. 5.5 have been shifted vertically to produce superposition at high molecular weights

centrations (155, 156). The data have been shifted by arbitrary amounts vertically (corresponding to arbitrary values of the friction factor ζ_0) to avoid overlap. Table 5.1 shows visually estimated values of cM at the intersection for the different concentrations. Figure 5.5 shows the same data, shifted vertically to achieve

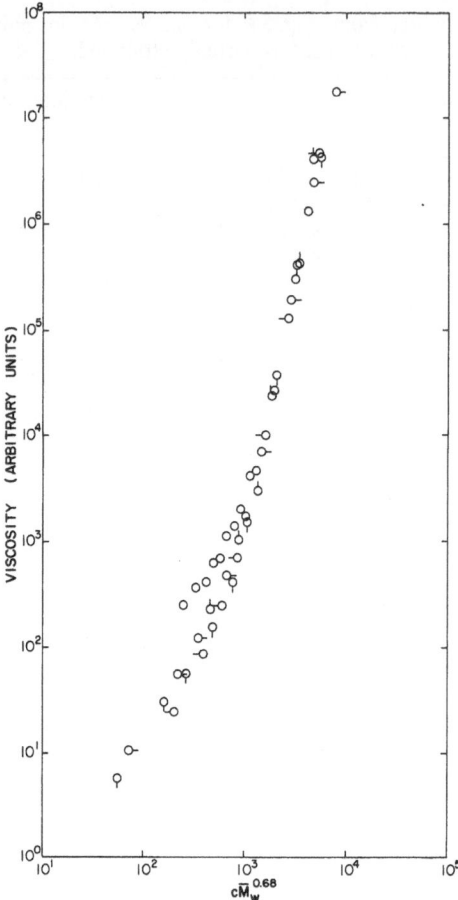

Fig. 5.6. Viscosity vs $c\bar{M}_w^{0.68}$ for polystyrenes at concentrations between 25% and 100%. The replotted data in Fig. 5.4 have been shifted vertically to produce superposition at high molecular weights

superposition at high molecular weights. The asymptotic lines are drawn with slopes 1.0 and 3.4.

Figure 5.6 shows the same data plotted as a function of $cM^{0.68}$ to test the low concentration reduction scheme based on $c[\eta]$ with a typical value of the Mark-Houwink exponent for good solvents. The data have been shifted vertically to achieve superposition at high molecular weights. It is clear that the cM variable produces a better superposition of data at all molecular weights and concentrations. The apparent variation in the values of cM at the intersections in Fig. 5.4 (Table 5.1) is largely due to a lack of data to define the limiting behavior at low molecular weights at some concentrations. The intersection on the superposed plot in the composite Fig. 5.5 is $cM = 30000$, giving $M_c = 30600$ for undiluted polystyrene ($\varrho = 0.98$ at $T = 217°$ C, in good agreement with the value 31200 reported by Berry and Fox (16).

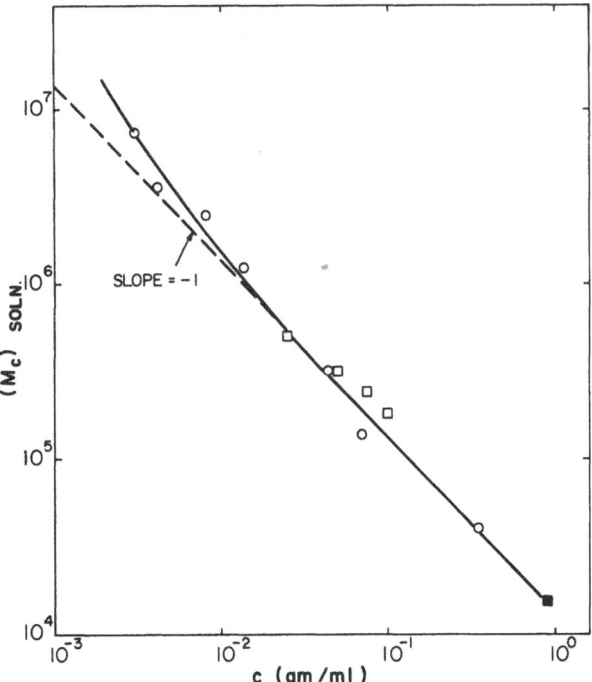

Fig. 5.7. Characteristic molecular weight for the viscosity vs concentration for the poly-isobutylene-toluene system. The open circles ($\bigcirc$) were obtained from η_0 vs c at constant molecular weight (159); the open squares ($\square$) were obtained from η_0 vs $\bar{M}_w$ at constant concentration (159); the solid square ($\blacksquare$) was obtained on undiluted polyisobutylene (9)

The thermodynamic properties of the diluent appear to be unimportant in determining $(M_c)_{\text{soln.}}$. Polyvinyl acetate in diethyl phthalate and in cetyl alcohol conform to Eq. (5.17), although the former is a good solvent and the latter a theta solvent at the temperature of measurement (157).

Establishment of $(M_c)_{\text{soln.}}$ becomes increasingly ambiguous at lower concentrations. The log-log slopes generally begin to differ from 1 and 3.4. In some cases the transition becomes more diffuse and sensitive to the nature of the solvent (132). In others, generally in polymers of low T_g, it may remain relatively sharp (158, 159) even at concentrations as low as 1–2% polymer. As noted earlier, the product $c[\eta]$ controls viscosity at low to moderate concentrations, such that one might expect behavior of the form $(M_c)_{\text{soln.}} \propto c^{-1/a}$ rather than $(M_c)_{\text{soln.}} \propto c^{-1}$. The data of Schurz and Hochberger on polyisobutylene-toluene solutions (159) are shown in Fig. 5.7. The values of $(M_c)_{\text{soln.}}$ at high concentrations obey Eq. (5.17), while at low concentrations there does appear to be a trend in the direction of $c^{-1/a}$ dependence.

This distinctive η_0 vs M behavior in melts and highly concentrated solutions is one of the features attributed to chain entanglement. The intensivity of M_c to temperature, the simplicity and insensitivity to solvent power of the concentration dependence of $(M_c)_{soln.}$, and the fact that M_c for different polymers corresponds to roughly the same number of main chain bonds, all suggest the onset of an intermolecular interaction based on geometrical rather than chemical attributes. It must be noted that the transition itself is not intrinsically a sharp one; indeed all sense of a transition is lost if the data are plotted in other ways, such as $\log \eta_0$ vs $M^{1/2}$ (9). The curves in the vicinity of the transition appear to be representable as the simple sum of high and low molecular weight forms (160):

$$\eta_0 = K c M \left[1 + \left(\frac{c M}{\varrho M_c} \right)^{2.4} \right].$$
(5.18)

The dashed line in Fig. 5.5 was calculated with this formula.

Rather than emphasizing the form of the transition region it is perhaps preferable to recall that the Rouse free-draining theory, with ζ_0 estimated independently from the diffusion coefficient of small molecules in the system (15) and without modification for interchain interactions, accounts almost quantitatively for the viscosity of concentrated systems at low molecular weights. For molecular weights beyond the neighborhood of $(M_c)_{soln.}$ the unmodified theory is not adequate. An intermolecular interaction of some sort, differing fundamentally from the $c[\eta]$ interactions and dependent on the product cM, is required to account for the behavior. The observed properties of M_c and $(M_c)_{soln.}$, cited above, are consistent with the type of interaction which might arise from a physical, rope-like coupling of chains, presuming that the number of couples along a molecule is proportional to the number of its intermolecular contacts.

Porter and Johnson (14) surveyed the literature prior to 1965, assembling M_c and $(M_c)_{soln.}$ data for a large number of polymers. Berry and Fox (16) have critically examined the values from this source and others, in many cases reanalyzing the data upon which the reported values were based. Apart from general questions about the accuracy of viscosity and molecular weight data, the main problem in determining M_c is the need to correct isothermal viscosity correlations to viscosities at constant segmental friction factor. Information is required on the temperature dependence of viscosity as a function of molecular weight, which is not always available. The corrections are largest for lower molecular weights, and are especially important if the viscosities are measured at temperatures within $50–100° C$ of T_g. Without correction, the slope of the low molecular weight branch of the curve is too large, and the deduced value of M_c is too big. Values of M_c for several well studied polymers are given in Table 5.2 (15, 16).

5.4.2. Properties of the Plateau Modulus

Other viscoelastic properties also acquire new characteristics at high concentration. In undiluted systems the long time end of the relaxation time distribution remains approximately Rouse-like for chains with molecular weight below M_c, as suggested by the agreement with reduced Rouse moduli in Fig. 5.2.

Table 5.2. Characteristic molecular weights for undiluted linear polymers

	M_e	M_c	M_c'
Polystyrene	18 100	31 200	130 000
Poly (α-methyl styrene)	13 500	28 000	104 000
1,4 Polybutadiene	1 900 (2 200)	5 900	13 800
Poly (vinyl acetate)	12 000	24 500	86 000
Poly (dimethyl siloxane)	8 100	24 400	61 000
Polyethylene	—	3 800	(14 400)
cis-Polyisoprene	5 800 (4 500)	10 000	60 000
Poly (methyl methacrylate)	5 900 (10 000)	27 500	>150 000
Polyisobutylene	8 900	15 200	—

The values of M_e [defined by Eq. (5.19)] are those reported by Ferry (*15*) and Berry and Fox (*16*), except that for polystyrene (*124*) and the values in parentheses for poly-methyl methacrylate (*146*), polybutadiene (*79, 197*), and cis-polyisoprene (*79, 197*). The values of M_c (defined as described in 5.4.1) are from the same compilations, except that for poly (α-methyl styrene) (*161*). The values of M_c' were obtained as described in Section 5.4.4.

As molecular weight increases, the slow relaxations shift progressively to longer times while the fast relaxations remain essentially unchanged.

This separation into two more or less distinct groups of relaxation processes is evident in master curves of the stress relaxation modulus (*162*), the shear compliance (*15, 163*), the dynamic moduli (*15, 124*), and the dynamic compliances (*15*). Figure 5.8 and 5.9 illustrate the behavior with master curves of $G'(\omega)$ and $G''(\omega)$ for undiluted polystyrene. At very high frequencies (not shown) $G'(\omega)$ is approximately 10^{10} dynes/cm^2 and comparatively constant (the glassy region). As ω decreases, $G'(\omega)$ begins to decrease (the transition region) in a manner which is independent of polymer molecular weight but which may differ considerably from one substance to another (*15*). The moduli of polymers and monomeric liquids (*164*) begin to differ qualitatively part way through the transition region ($G'(\omega) \sim 10^8$ dynes/cm^2). The former decay more slowly and in many polymers appear to approach the form dictated by the Rouse spectrum, $G'(\omega) \propto \omega^{1/2}$. The high frequency slopes are approximately 1/2 in Figs. 5.8 and 5.9.

For long chains, the modulus at about 10^6–10^7 dynes/cm^2 passes into a domain of much slower relaxation (the plateau region). The width of the plateau depends strongly on molecular weight, while the characteristic modulus of the plateau region G_N° is independent of molecular weight. The presence of the plateau confers rubber-like elastic properties over an intermediate range of frequencies or times.

Although $G'(\omega)$ appears to be relatively flat on log-log plots, especially for very high molecular weights and narrow molecular weight distributions, the plateau region is not entirely devoid of relaxation processes. This is most easily seen in $G''(\omega)$ (Fig. 5.10) which passes through a shallow minimum in the plateau but clearly does not drop to zero. The plateau occupies the span of time between the transition region and the terminal region, and those two groups of relaxations always appear to overlap to some extent. In order to evaluate G_N°, which may be defined as the modulus associated with the terminal relaxation processes only,

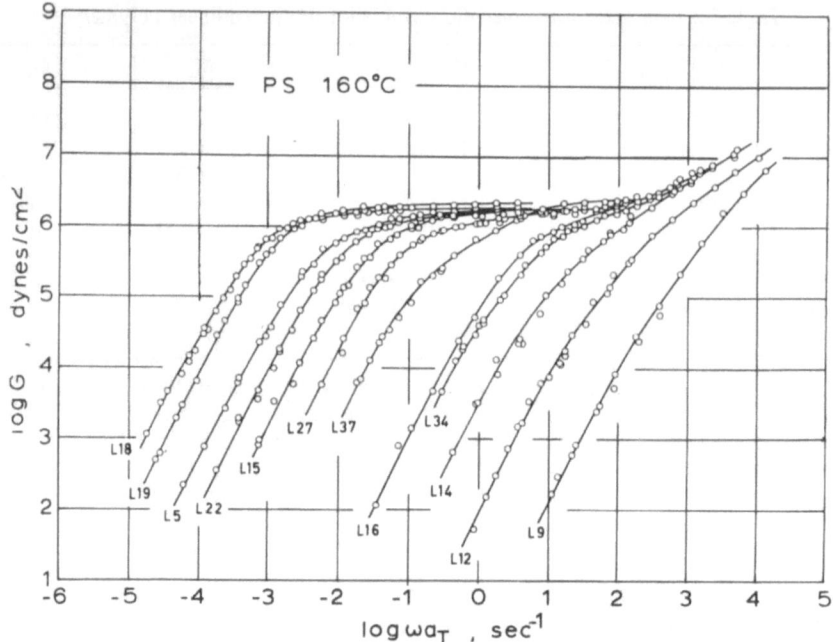

Fig. 5.8. Storage modulus vs frequency for narrow distribution polystyrene melts, reduced to 160° C by temperature-frequency superposition. Molecular weight range from $\bar{M}_w = 8900$ (L9) to $\bar{M}_w = 581000$ (L18) (*124*). [Reproduced from Macromolecules **3**, 111 (1970).]

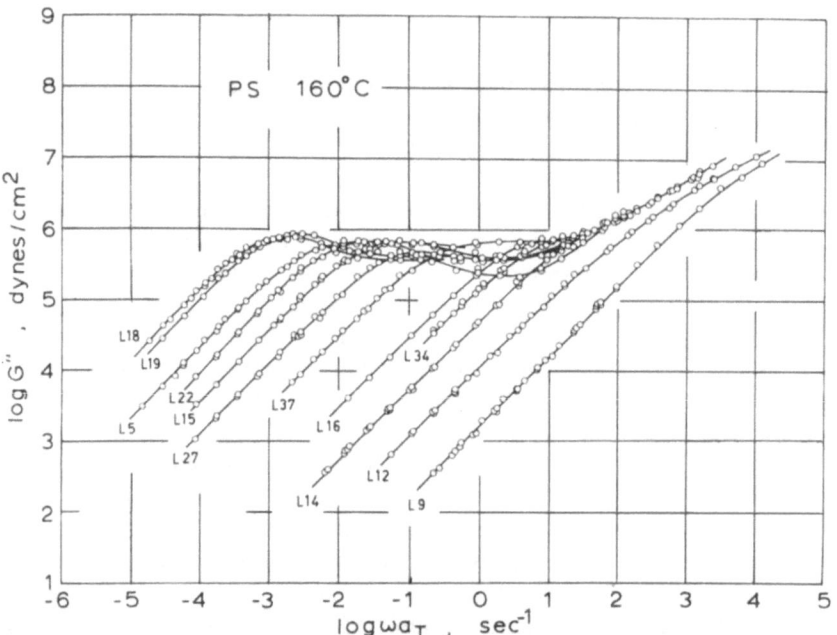

Fig. 5.9. Loss modulus vs frequency for narrow distribution polystyrene melts (*124*). See Fig. 5.8. [Reproduced from Macromolecules **3**, 112 (1970).]

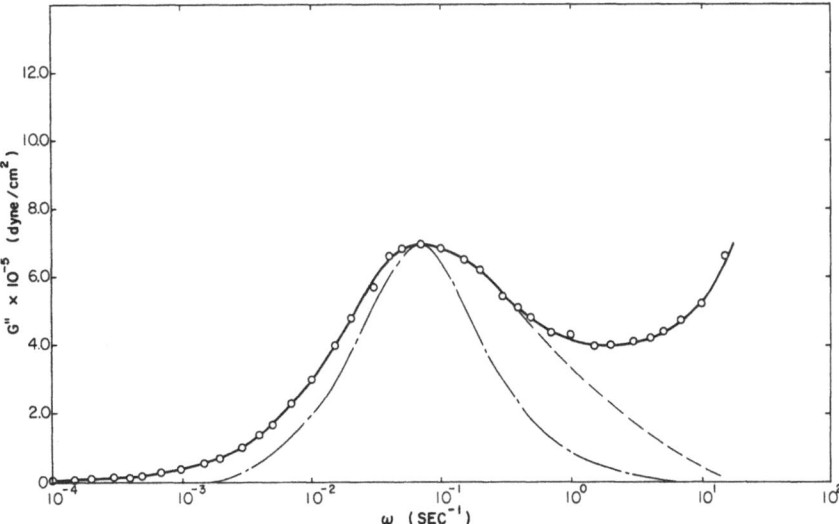

Fig. 5.10. Loss modulus vs frequency for a narrow distribution polystyrene melt reduced to 160° C ($\bar{M}_w = 215000$) (124). The dashed line (– – –) is an approximate resolution of the terminal relaxation peak. The construction line (– · –) is G'' vs ω calculated for a process with a single relaxation time $[G''_{(\omega)}/G''_{\max} = 2\omega\tau/(1 + \omega^2\tau^2)]$

one must separate the two contributions. Unfortunately, there is no unambiguous way to make the separation. If a form is selected for the terminal relaxation behavior, G_N° can be extracted directly (165). A common method is to choose the value of the modulus at the inflection point of $G(t)$. Creep compliance data can be handled similarly (166), the plateau modulus G_N° being then simply the reciprocal of the plateau compliance J_N°. Other workers prefer to convert to relaxation spectra, resolve the transition and terminal regions, and calculate J_N° or G_N° from the area under the terminal peak (167).

Oser and Marvin (168) have developed a phenomenological theory involving two groups of relaxations, a Rouse spectrum at short times and a second, shifted Rouse spectrum at long times. They show that certain properties of $G''(\omega)$ and $J''(\omega)$, the frequencies of their minima in the plateau region, and the heights and frequencies of their terminal loss peak maxima, are related in simple ways to G_N° and η_0. Ferry and co-workers determine G_N° from the area under the terminal peak in $J''(\omega)$ through Eq. (3.29). The problem of resolving overlap in the terminal and transition peaks always exists, but for narrow distribution samples of high molecular weight the separation can be handled in a reasonably objective manner. Using $J''(\omega)$, Ferry and co-workers have evaluated G_N° for a large number of rubbery polymers (15) and have shown that the Marvin-Oser equations, after slight adjustments in their numerical coefficients, are able to give consistent estimates also. Onogi and co-workers have used the area under the $G''(\omega)$ loss peak to estimate G_N° in polystyrene (124) and polymethyl methacrylate (146).

The plateau region appears when the molecular weight exceeds $M_c[(M_c)_{soln.}$ for solutions], and is taken to be a direct indication of chain entanglement. Indeed the presence of a plateau may be a more reliable criterion than η_0 vs M behavior, especially in solutions of moderate concentration where viscosity may exhibit quite complex concentration and molecular weight behavior. It is postulated that when M greatly exceeds M_c, a temporary network structure exists due to rope-like interlooping of the chains. Rubber-like response to rapid deformations is obtained because the strands between coupling points can adjust rapidly, while considerably more time is required for entire molecules to slip around one another's contours and allow flow or the completion of stress relaxation.

Plateau moduli are commonly expressed in terms of M_e, the *apparent* average molecular weight between coupling junctions. The modulus equation from the kinetic theory of rubber elasticity $\left(\text{Eq. (7.2) with } g\,\dfrac{\langle r^2 \rangle}{\langle r^2 \rangle_0} = 1 \text{ and } v = c/M_e \text{ or}\right.$ $\left. \varrho/M_e \text{ for undiluted polymers}\right)$ is used, giving

$$M_e = \frac{\varrho RT}{G_N^\circ}. \tag{5.19}$$

The number of junction points per molecule is therefore $E = M/M_e - 1$. Values of M_e for a number of polymers have been tabulated recently by Ferry (15) and Berry and Fox (141). Selected values are given in Table 5.2. The close parallel in the values of M_e and M_c is apparent.

In concentrated solutions, the value of $(M_e)_{soln.}$ is inversely proportional to concentration, such that:

$$(G_N^\circ)_{soln.} = \left(\frac{c}{\varrho}\right)^2 G_N^\circ, \tag{5.20}$$

or

$$(M_e)_{soln.} = \left(\frac{\varrho}{c}\right) M_e. \tag{5.21}$$

This behavior, again paralleling that of $(M_c)_{soln.}$, is now clearly established for a number of polymers (15, 142, 166, 167). Figure 5.11 shows Eq. (5.21) to be valid over approximately the same range of concentrations as the corresponding equation for $(M_c)_{soln.}$.

5.4.3. Properties of the Terminal Zone Spectrum

Relaxation time spacings in the terminal region are quite different than the Rouse spacings: much more tightly grouped around the mean but with a tail of faster relaxations which trails back into the plateau region. Tobolsky and co-workers have found that the empirical form $G(t) = A \exp[-(t/r)^B]$ fits experimental stress relaxation data in the terminal region (165). The parameter B correlates with molecular weight distribution with values of ca. 0.60 for typical narrow distribution polystyrenes and an extrapolated value $B = 0.75$ for hypothetical

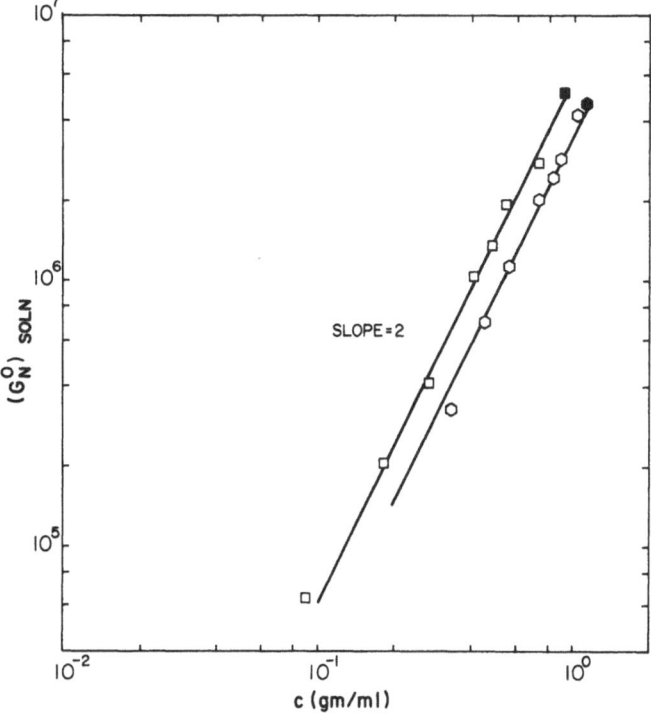

Fig. 5.11. Plateau modulus vs. concentration for polymethyl methacrylate $\Diamond$ *(142)* and cis-polyisoprene $\square$ *(167)*. The solid points are undiluted samples

monodisperse specimens. The dynamic moduli and the molecular weight dependence of steady-state shear compliance (Section 5.4.4) demonstrate the strikingly sharp character of the terminal spectrum (compared to the Rouse spectrum) for narrow distribution polymers.

Terminal relaxations dominate the steady flow behavior. The zero-shear viscosity has already been discussed. For long chains the steady-state compliance, J_e°, and the characteristic relaxation time of the terminal region τ_m (as determined for example by the various procedures of Tobolsky and co-workers *(165, 169)* and Ferry and co-workers *(123)* depend on molecular weight and concentration in a manner quite different from that given by the corresponding Rouse forms *(169, 173)*. Despite these differences, the observed orders of magnitude of τ_m and J_e° are surprisingly similar to values calculated from the Rouse theory *(11, 170)*. For $M \gg (M_c)_{\mathrm{soln.}}$, the essential features are:

Observed form	*Rouse form*

$$\tau_m \propto \frac{\eta_0}{c^b T} \qquad\qquad \tau_m = \frac{6}{\pi^2} \frac{\eta_0 M}{cRT}$$

$$J_e^\circ \propto \frac{1}{c^b T} \qquad\qquad J_e^\circ = \frac{2}{5} \frac{M}{cRT}$$

The exponent b is a matter of some current dispute, being 2 in nearly all studies but appearing to range as high as 3 in a few instances (Section 5.4.4). In any case, the explicit molecular weight dependence is lost, and concomitantly the concentration dependence is increased (*173*). If all relaxation times had been scaled up by the same factor (the relative spacings and intensities of the Rouse spectrum being retained), the Rouse forms would have been preserved for $M \gg (M_c)_{soln}$.

5.4.4. The Steady State Compliance

Values of J_e° are obtainable from several types of experiments (Table 5.3). In principle, all the methods should agree, subject to very broad assumptions (Section 3). However, each presents its own set of experimental problems and sources of systematic error. Thus, values of J_e° obtained from stress relaxation are very sensitive to the shape of $G(t)$ for long times, the region where the residual stress itself is extremely small and difficult to measure accurately. Values obtained from the intercept of the creep curve and from creep recovery are sensitive to the attainment of steady state in the flow. Of the two, creep recovery is clearly preferred because elastic recoil and hence J_e° is measured directly. For all methods based on linear viscoelasticity the stresses of course must be small enough to place the response in the linear viscoelastic region. Values from the normal stress function are usually determined from the total axial thrust developed in plate-cone viscometers. Deviations from a quadratic dependence on shear rate are usually beginning to appear in the experimentally accessible region, and a somewhat uncertain extrapolation to zero shear rate is required. Plots of N_1 vs σ^2 are usually much more linear, but still there remains the possibility of changes at

Table 5.3

Method for obtaining J_e°	Equation from phenomenological viscoelasticity
Stress relaxation	$\lim\limits_{t \to \infty} \int\limits_0^t \xi G(\xi)d\xi$
Recoverable shear after steady-state deformation	$\lim\limits_{t \to \infty} [\gamma_r/\sigma_0]$
Intercept of steady-state creep curve	$\lim\limits_{t \to \infty} \{J(t) - t/\eta_0\}$
Steady-state dynamic response	$\dfrac{1}{\eta_0^2} \lim\limits_{\omega \to 0} \left\{ \dfrac{G'(\omega)}{\omega^2} \right\}$
Normal stress measurements	$\dfrac{1}{2\eta_0^2} \lim\limits_{\dot\gamma \to 0} \left\{ \dfrac{N_1}{\dot\gamma^2} \right\}$
Flow birefringence	$\dfrac{1}{\eta_0} \lim\limits_{\dot\gamma \to 0} \left[\dfrac{\cot 2\chi}{\dot\gamma} \right]$

lower shear stresses. Values from steady flow birefringence rely in addition on the validity of the stress-optical law (18).

The data collected for review here were obtained on samples of narrow, but still somewhat variable, molecular weight distribution. Steady-state compliance is highly sensitive to polydispersity, so some disagreement among investigators is to be expected, due simply to slight differences in the distributions of their samples. The difficulties in all methods are probably somewhat reduced in systems of low polydispersity because the distribution of relaxation times is narrow, and relatively high stresses can be imposed before non-linearities appear. Nevertheless, the presence of even small amounts of high molecular weight species can increase J_e^0 significantly (163). Non-linear response is difficult to avoid in such circumstances and may cause apparently different values to result from the various methods.

The steady-state compliance data have been analyzed in reduced form

$$J_{eR} = \frac{J_e^0 \, cRT}{M} \left(\frac{\eta_0}{\eta_0 - \eta_s} \right)^2 .$$

(5.22)

This form for reduced compliance is suggested by the dilute solution molecular theories, according to which J_{eR} is governed by the dispersity of molecular relaxation times [from Eq. (4.12)]:

$$J_{eR} = \frac{\sum\limits_N \tau_i^2}{\left(\sum\limits_N \tau_i \right)^2} .$$

(5.22a)

Variations in J_{eR} at finite concentration reflect the influence of intermolecular interaction on the *relative* spacings of the long relaxation times. Through J_{eR} these effects may be examined separately from effects on the magnitude of the viscosity (or the average magnitude of the long relaxation times). The ratio $\sum \tau_i^2 / (\sum \tau_i)^2$ is called S_2/S_1^2 by Ferry and co-workers. The above expression for J_{eR} reduces to their $J_{eR}^0 = \dfrac{RT}{M[\eta]^2} \lim\limits_{c \to 0} \dfrac{J_e^0}{c}$ at zero concentration. It becomes $J_e^0 cRT/M$ when $\eta_0 \gg \eta_s$, which is the definition of reduced compliance used by Tschoegl (183) and Janeschitz-Kriegel (18). For the Rouse model $J_{eR} = 0.4$; the Zimm model gives $J_{eR} = 0.206$.

Figure 5.12 shows J_{eR} obtained by different investigators on a single sample of narrow distribution polystyrene [sample 6a, $\bar{M}_w = 860.000$, $\bar{M}_w/\bar{M}_n = 1.05$ (174), polymerized anionically by the Pressure Chemical Company, Pittsburgh, Pennsylvania]. A number of different solvents, temperatures, and methods are represented (175–182). The values rise from below 0.4 at low concentrations, pass through a broad maximum which somewhat exceeds 0.4, then decrease again at high concentrations. Data from dynamic moduli, normal stress measurements, and

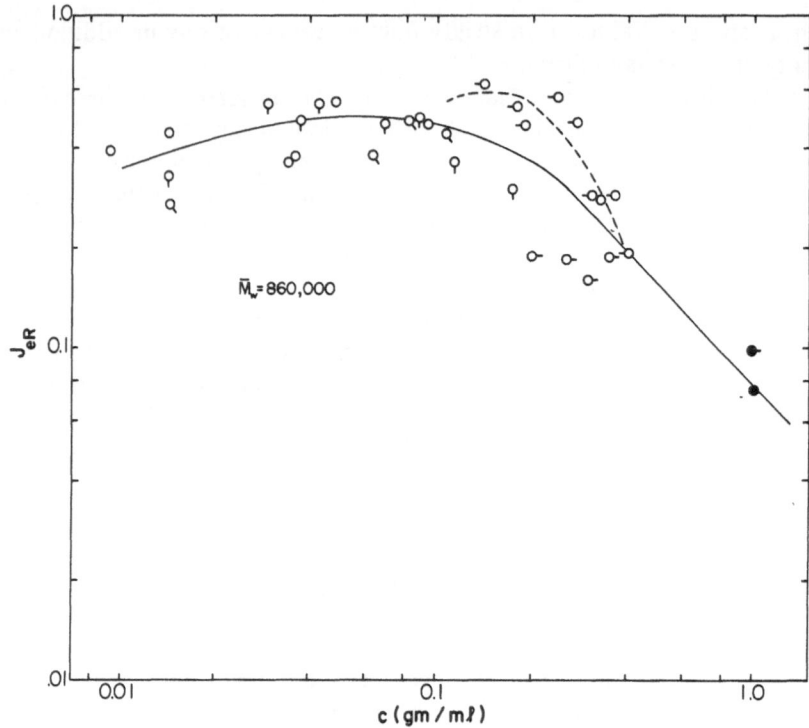

Fig. 5.12. Values of reduced compliance J_{eR} for solutions of a narrow distribution poly-styrene ($\bar{M}_w = 860000$) according to several investigators and methods. Symbols: $\male$ from $G'(\omega)$ (*175, 176*), $\leftarrow$ from N_1 (*177*), $\circ\!\!-$ from N_1 (*178*), $\circ$ from flow birefringence (*179*), $\bullet\!\!-$ from flow birefringence (*180*), $\bullet$ from $G'(\omega)$ and N_1 (*181*), and $-\!\!\circ$ from creep recovery (*182*)

flow birefringence seem consistent with each other and generally follow the same trend. Values from creep recovery differ (*182*), exceeding those from other methods by approximately a factor of two in the concentration range 0.15–0.25 gm/ml, then decreasing rapidly towards them again at higher concentrations. Comparison of literature data on two other Pressure Chemical samples [sample 3a, $\bar{M}_w = 411,000$ in Fig. 5.13 (*177–179, 181, 182, 184, 185*), and sample 14a, $\bar{M}_w = 1,800,000$] shows similar behavior. The reported creep recovery results (*182*) exceed other values for J_e^0 at intermediate concentrations by factors as large as three.

This discrepancy is rather disturbing because the creep recovery experiments were obviously designed and tested with considerable care (*186*). Furthermore, as pointed out earlier, creep recovery is the most direct method for measuring J_e°. Most of the likely errors (such as inertial effects in the instrument or non-attainment of steady state) would tend to give values which are too small rather than too large. Similar but smaller effects have been observed by other methods in solutions of polyisoprene (*167*) and poly (α-methyl styrene) (*187*). In the latter

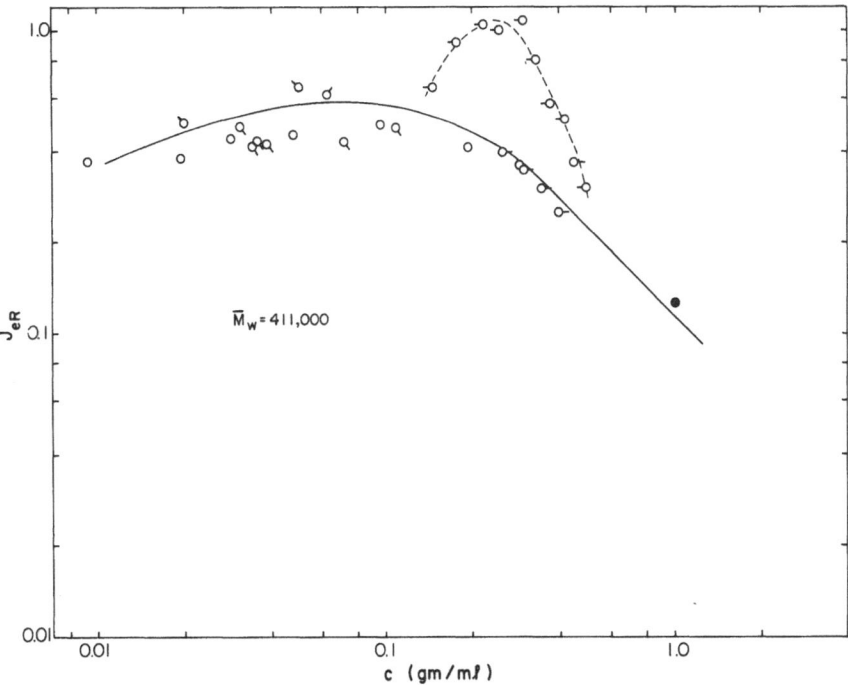

$\bar{M}_w = 411,000$

Fig. 5.13. Values of J_{eR} for solutions of a narrow distribution polystyrene ($\bar{M}_w = 411000$). According to several investigators and methods. Symbols are the same as Fig. 5.12, except $\circ$ from N_1 (184) and $\circ$ from three methods (185)

a correspondence was noted with a jelly-like appearance of the solution. It is entirely possible that creep recovery is detecting a real effect in the region where reduced compliance is beginning to fall, one which is missed or seen only imperfectly by the less delicate methods. Recognizing this possibility, we will nevertheless assume that normal behavior corresponds to the general trend of data in Figs. 5.12 and 5.13 and omit the data in Ref. (182) from later correlations.

Figure 5.14 shows J_{eR} for undiluted, narrow distribution polystyrenes from a variety of investigators and methods (124, 163, 165, 180, 181, 188, 189, 191–193). Values from one study using stress relaxation (188) lie below the general trend. However, stress relaxation data on the same samples (the Dow S series polystyrenes), when first fitted to an analytical form for $E(t)$, yield values in good agreement with the other data in Fig. 5.14 (165). It seems clear that the stress relaxation method is sensitive to extrapolation errors. Values from a study using dynamic moduli appear to be consistently higher than the general trend for molecular weights less than 10^5 (124). Values obtained from creep recovery (163), normal stresses (181, 193), and flow birefringence (180) are in fairly good agreement; values from tensile creep (191), the dynamic moduli (124, 192), and

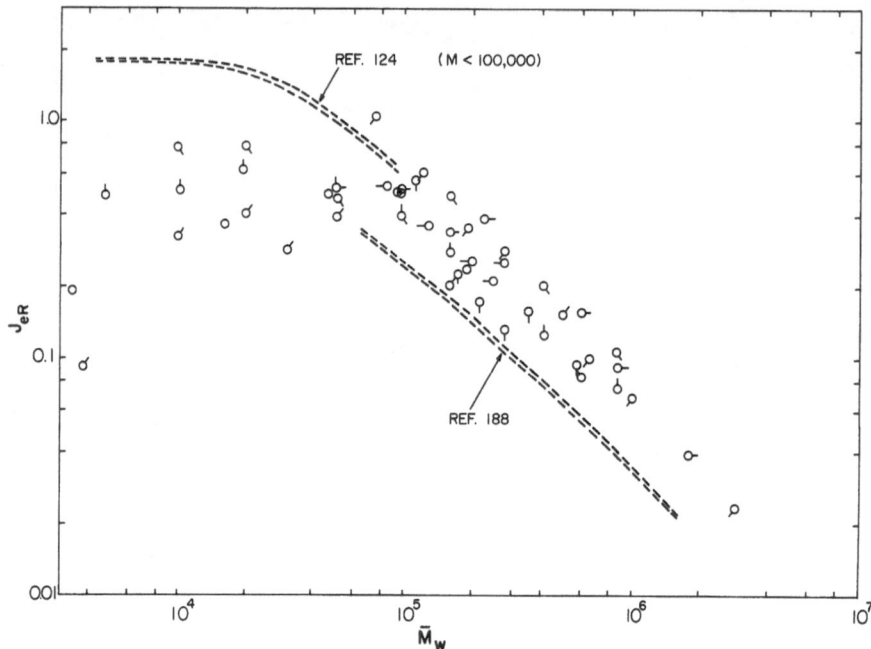

Fig. 5.14. Reduced compliance vs molecular weight for undiluted polystyrenes of narrow molecular weight distributions. Symbols are: $\bigcirc$ from creep recovery *(163)*, $\bigcirc\!\!\!\!^{\prime}$ from $G'(\omega)$ *(192)*, $\bigcirc\!\!-$ from flow birefringence *(180)*, $\bigcirc\!\!\!\!\diagdown$ from N_1 *(189)*, $\bigcirc\!\!\!\!\vert$ from $G'(\omega)$ ($M > 10^5$ only) *(124)*, $\bigcirc\!\!\!\!\diagdown$ extrapolated from steady state creep *(191)*, $-\bigcirc$ from stress relaxation *(165)*, and $-\bigcirc\!\!\!-$ from N_1 *(193)*

Table 5.4. Values of J_{eR} for narrow distribution polystyrenes from Pressure Chemical Company, Pittsburgh, Pa., USA; $J_{eR} = J_e^\circ \varrho RT / \bar{M}_w$

M_w	Prest *(181)*	Mills and Wales *(190)*	Nevin *(192)*	Mieras and van Rijn *(189)*	Crawley *(193)*
10 200	0.51˙		0.32	0.77	
19 700	0.62		0.40	0.77	
51 000	0.52	0.53	0.39	0.47	
97 000	0.40	0.51	0.48	0.40	0.47
160 000	0.28	0.34	0.20	0.49	0.24
411 000	0.13			0.20	
860 000	0.075	0.092		0.11	

some of the normal stress data *(189)* are slightly displaced. The values of Zosel *(194)*, after his corrections for polydispersity, are also in good agreement.

The scatter may perhaps be partially accounted for by differences in polydispersity among the samples. However, as Table 5.4 shows, even measurements on the same polymer may result in values which differ appreciable among investigators. On the whole, the overall behavior is fairly clear: J_{eR} maintains a value near but perhaps slightly above the Rouse value for $10^4 < M < 10^5$, then becomes approximately proportional to $1/M$ for higher molecular weights[3].

Figure 5.15 shows reduced compliance for polystyrene solutions from several sources (175–179, 184, 185), combined and plotted as a function of cM. All are narrow distribution samples. Selected melt data from Fig. 14 are also included (163, 180, 181). The preponderance of data are in fair accord with the Rouse value for $5000 \gtrsim cM \gtrsim 100\,000$. As in the case of the melt data, they tend to rise slightly above 0.4, influenced perhaps by residual polydispersity in all the samples. The shallow maximum in this region may well be caused also by the fact that only a few of the longest relaxation times are shifted at the smaller values of cM, as suggested by Janeschitz-Kriegl (18). The reduced compliances tend to be smaller and more scattered at still smaller values of cM, probably reflecting an approach to Zimm-like behavior for high molecular weight samples at low concentrations, and the loss of flexible coil character for low molecular weight samples at high concentrations (15). Also, as noted earlier, $c[\eta]$ rather than cM is the appropriate reducing variable for J_{eR} at low and moderate concentrations. For values of cM greater than about $130\,000$, the reduced compliance acquires a dependence on cM, becoming proportional to $(cM)^{-1}$. No important difference in reduced behavior between solutions and melts is apparent despite the $100–200°$ C difference in the temperatures of measurement.

Confirming an earlier conjecture (195), the data on several linear polymers follow a similar pattern. Figures 5.16 and 5.17 contain results for narrow distribution samples of 1,4 polyisoprene (166, 167, 196, 197) and poly (α-methyl styrene) (161, 176, 198–201). The onset of $(cM)^{-1}$ behavior seems clearly related to the same processes that produce the plateau region and the transition in η_0 vs M behavior. In almost all cases $(cM)^{-1}$ dependence in J_{eR} begins at a

---

[3] The steady-state compliance is exceedingly sensitive to the high molecular weight tail of the molecular weight distribution. Evaluation of J_e° depends on a satisfactory handling of two limiting processes. The stress or deformation must be small enough to achieve linear viscoelastic response, and the time must be long enough (in the case of creep and creep recovery) or the frequency low enough (in the case of dynamic experiments) to reach the limiting response. These conditions merge in the normal stress and flow birefringence methods, requiring low enough stress to achieve the limiting form, with no further changes at still lower stresses. The contribution of trace amounts of high molecular weight components in an otherwise narrow distribution sample is very sensitive to stress. It is possible, although it is only a conjecture, that this contribution may be lost in methods such as normal stress measurements which involve relatively high stresses and large deformations. The same sample in low stress methods such as creep recovery may then yield higher values of J_e° unless the tails have been carefully removed by fractionation. From the standpoint of obtaining the true value of J_e° for a sample therefore, the creep recovery is clearly preferred.

However, to obtain a value which represents the contribution of everything except a trace tail, the normal stress measurement may in fact give acceptable results. It is somewhat reassuring that J_e° values deduced from normal stress and flow-birefringence data on undiluted narrow distribution polystyrenes agree with the Plazek results (163), the latter having been obtained by creep recovery with very carefully fractionated samples. Similarly, the high values of J_e° obtained by creep recovery on solutions of the Pressure Chemical Samples (182) may be the result of trace tails at high molecular weight whose contribution is missed by the other methods. It is obviously dangerous to place too much reliance on such fortuitous behavior, however.

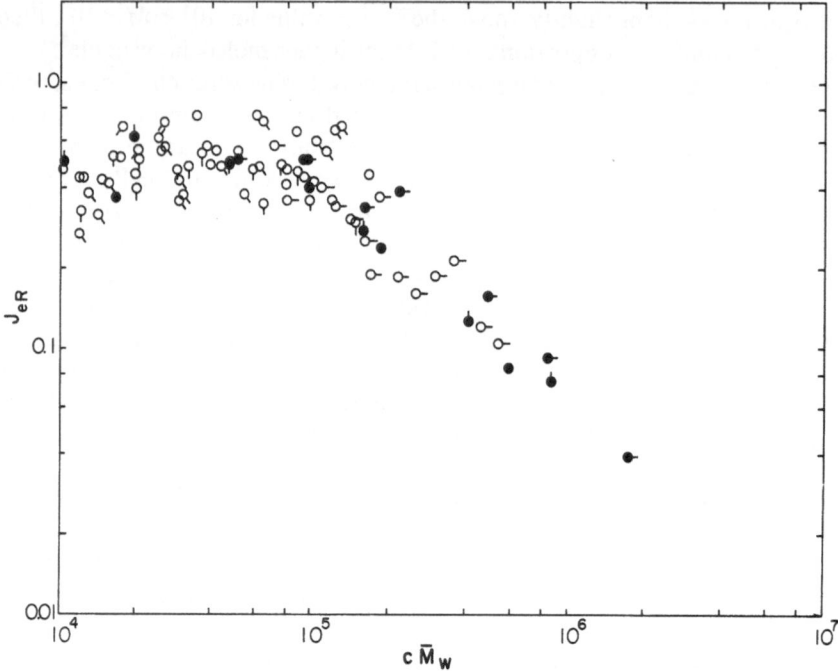

Fig. 5.15. Reduced compliance vs the $c\bar{M}_w$ product for solutions and undiluted samples of narrow distribution polystyrenes. Symbols are: ○ from flow birefringence (*179*), ◌ from several methods (*185*), ○- from N_1 (*178*), ◌ from $G'(\omega)$ and N_1 (*177*), ○ from $G'(\omega)$ (*175*, *176*), ○ from N_1 (*184*), ● from creep recovery (*163*), ● from $G'(\omega)$ and N_1 (*181*), and ●- from flow birefringence (*180*). The filled points are undiluted polystyrenes

characteristic value of cM which is roughly 3–4 times the value of $\varrho\bar{M}_c$. Figure 5.18 is a combined plot of reduced compliance vs $cM/\varrho M_c$, including fragmentary data on narrow distribution polyvinyl acetate (*176*, *195*), polybutadiene (*202*, *203*), and poly(dimethyl siloxane) (*204*). A rather tenuous estimate for polyethylene (*204*) has also been included. Some published data on narrow distribution polyvinyl acetate (*205*, *206*) have been omitted because of possible complications due to long chain branching.

The meaning of these observations on narrow distribution systems is the following (*153*). In undiluted polymers the steady-state compliance is directly proportional to molecular weight and approximately equal to the Rouse value for molecular weights both large enough to display random coil behavior, but smaller than a characteristic molecular weight M_c'. For molecular weights greater than M_c' the compliance approaches a constant value $(J_e^0)_\infty$. Values of $(J_e^0)_\infty$ estimated for the polymers in Fig. 5.18 are given in Table 5.5. Values for truly monodisperse samples would undoubtedly be slightly smaller. The characteristic molecular weight M_c' is a small multiple of M_c. The same molecular weight behavior occurs in concentrated solutions and the characteristic value, $(M_c')_\text{soln.}$, has the same concentration dependence as $(M_e)_\text{soln.}$ and $(M_c)_\text{soln.}$, namely,

$$(M_c')_\text{soln.} = \frac{\varrho}{c} M_c'. \tag{5.23}$$

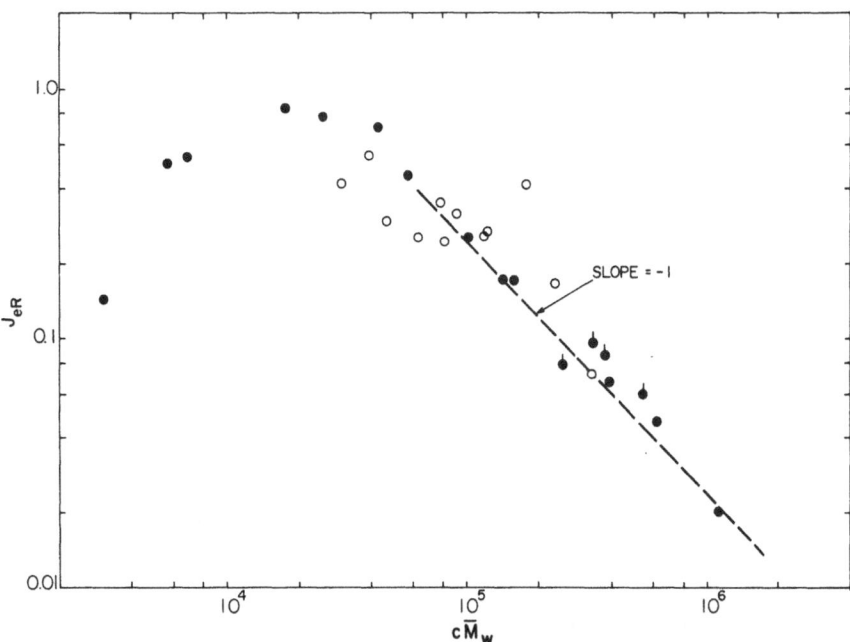

Fig. 5.16. Reduced compliance vs $c\bar{M}_w$ for solutions and undiluted samples of cis-polyisoprene. Symbols are: ● undiluted samples from steady state creep (*166, 196*), ◓ undiluted samples from N_1 (*197*), and ○ solutions extrapolated from steady state creep (*196*)

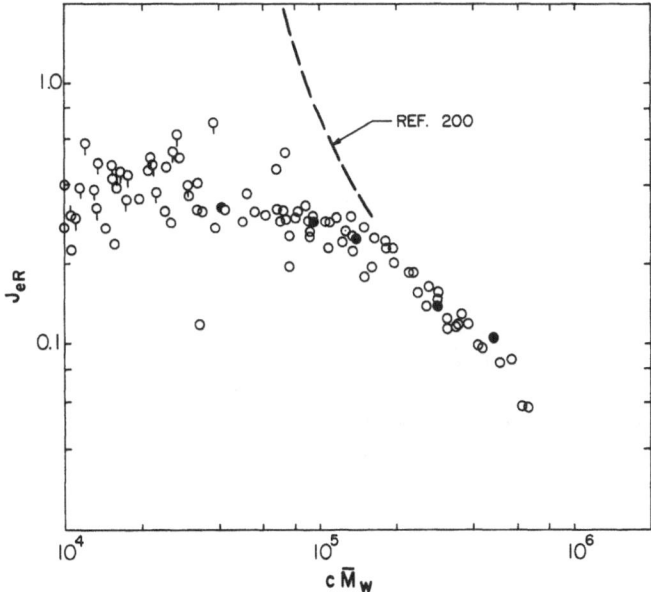

Fig. 5.17. Reduced compliance vs $c\bar{M}_w$ for solutions and undiluted samples of poly (α-methyl styrene). Symbols are: ○ from N_1 (*198, 199*), ◔ from $G'(\omega)$ (*176*), ♀ from N_1 (*201*), and ● from stress relaxation (*161*)

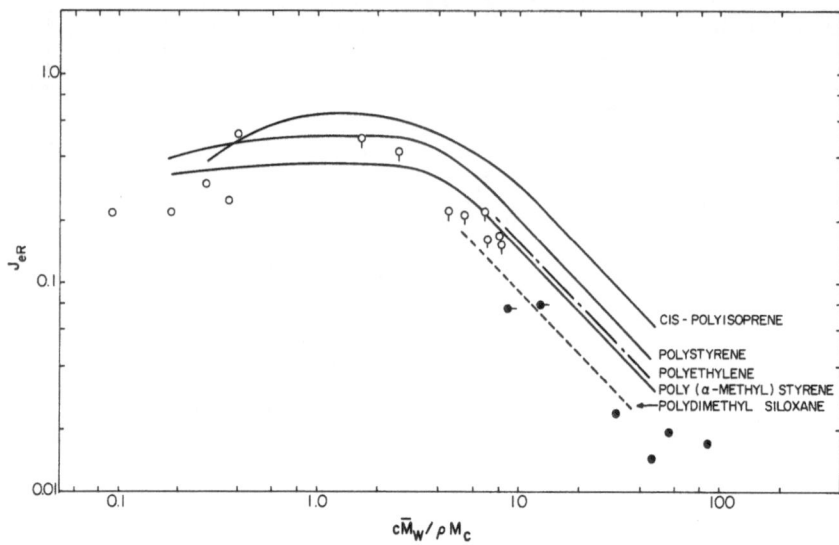

Fig. 5.18. Reduced compliance vs $c\bar{M}_w/\varrho M_c$ for solutions and undiluted sample of linear, narrow distribution polymers. The lines for polyethylene and (polydimethyl siloxane) are based on $G'(\omega)$ data as reported by Mills (204). Points are selected values for polyvinyl acetate solutions, ○ (176) and ♀ (195), and undiluted 1,4 polybutadiene, ● (202) and ●- (203) The values of M_c were taken from Table 5.2

Table 5.5. Limiting compliance for undiluted linear polymers of narrow distribution

Polymer	$T°$ C	$(J_e^0)_\infty$ (cm²/dyne)	G_N^0 (dyne/cm²)	$G_N^0(J_e^0)_\infty$
Polyethylene	190	0.22×10^{-6}	—	—
1,4 Polybutadiene	25	0.25	11.5×10^6	2.9
Poly(α-methyl styrene)	200	1.0	3.2	3.2
Poly(dimethyl siloxane)	20	1.0	3.0	3.0
Polyvinyl acetate	30	1.2_5	2.6_5	3.3
cis-1,4 Polyisoprene	−30	1.4	3.9	5.5
Polystyrene	200	1.75	2.0	3.5

The steady-state compliance is, furthermore, inversely proportional to the first power of concentration and nearly equal to the Rouse value for concentrations high enough for appreciable coil overlap to occur but less than $\varrho M_c'/M_c'$. At higher concentrations the compliance becomes inversely proportional to the square of concentration. Differences in the reduced compliance due to temperature and solvent power are negligible, and differences from one polymer system to another are greatly reduced if the polymers are compared through the reduced variable $cM/\varrho M_c$. Values of M_c' were obtained from the intersection of lines drawn through data at large and small values of cM and are given in Table 5.2. When data were available only in the asymptotic region for large values of M, as in the case of 1,4-polybutadiene, polyethylene and poly(dimethylsiloxane), values of M_c' were estimated for the undiluted polymer with the relation for the intersection with Rouse behavior: $M_c' = (J_e^0)_\infty \varrho RT/0.4$. Judged by the agreement of data on even high molecular weight fractions of polymethyl methacrylate with the Rouse theory (146), the value of M_c' for that polymer exceeds 150000.

One early proposal for the form of the monodisperse compliance curve, based primarily on data beyond $(M_c')_{soln.}$, was (178):

$$\frac{1}{J_{eR}} = \alpha_1 + \alpha_2 \frac{cM}{\varrho M_c} \tag{5.24}$$

in which the parameters α_1 and α_2 were taken to be independent of c, M, and T, and were assumed to be the same for all linear polymers. Another proposal, cast in a similar dimensionless form, is (207):

$$\left(\frac{1}{J_{eR}}\right)^2 = \alpha_1^2 + \alpha_2^2 \left(\frac{cM}{\varrho M_c}\right)^2 \tag{5.25}$$

The subsequent comparisons of several authors (153, 199, 208) have shown that the change in behavior near $(M_c')_{soln.}$ is more abrupt than the rather gradual transition implied by Eq. (5.24). Equation (5.25) gives a behavior somewhat closer to observations. The combined results in Fig. 5.18 can be summarized by the expression (for $\eta_0 \gg \eta_s$):

$$J_e^0 = \frac{0.4\,M/cRT}{\left[1 + 0.08\left(\frac{cM}{\varrho M_c}\right)^2\right]^{1/2}}. \tag{5.26}$$

Because of apparent residual variations in the ratio M_c'/M_c among the different polymers, a more accurate expression can be formed with the reduced variable $cM/\varrho M_c'$:

$$J_e^0 = \frac{0.4\,M/cRT}{\left[1 + \left(\frac{cM}{\varrho M_c'}\right)^2\right]^{1/2}} \tag{5.27}$$

Again the reader must be warned that a large proportion of the J_e^0 data in the correlation, summarized in Eqs. (5.26) and (5.27) and Tables 5.2 and 5.5, are based on normal stress measurements (total thrust in plate-cone rheometers) with attendant uncertainties about whether limiting behavior was attained. Also, in some polymers, such as polyethylene, polyvinyl acetate, polydimethyl siloxane, and polybutadiene, they are based on the results of only a few investigators or on a limited number of samples with narrow but uncertain polydispersity. Even in the most thoroughly studied polymer, polystyrene, there are significant unresolved questions on the concentration dependence of J_e^0 and even on the value of J_e^0 for particular samples. Thus, although Eqs.(5.26) and (5.27) are attractively simple, and probably describe overall behavior fairly well, one must expect some variations from these forms as more complete and accurate data appear.

5.4.5. Effects of Molecular Weight Distribution

The discussion to this point has dealt with polymers of narrow molecular weight distribution. Distribution, in fact, affects all viscoelastic properties of the plateau and terminal regions (15). As an molecular weight distribution broadens, the relaxation time distribution broadens in a more or less corresponding manner. Increasing amounts of relaxation occur throughout the plateau region, making it progressively less well defined. Two plateau regions may be discerned in binary blends with a sufficiently large difference in the component molecular weights (209). The initial plateau modulus G_N^0 appears to be independent of molecular weight distribution, subject of course to uncertainties due to the increasing diffuseness of the plateau with increasing polydispersity.

Studies on both binary blends and whole polymers of moderate polydispersity have established that η_0 at constant friction factor ζ_0 depends on the weight-average molecular weight alone (15, 16, 181, 209):

$$\eta_0 = \eta_0(\bar{M}_w) \tag{5.28}$$

in which $\eta_0(M)$ is the viscosity for a monodisperse polymer of molecular weight M. Thus, even above M_c, this characteristic of the Rouse model [Eq.(4.27)] is borne out. Exceptions *may* exist in samples of very broad distribution, where some data suggest an average lying between $\bar{M}_w$ and $\bar{M}_z$ (210, 210a).

The recoverable compliance J_e^0 is very sensitive to molecular weight distribution, especially to the tail of the distribution at high molecular weights. According to the Rouse model [Eq.(4.28)], when samples with the same $\bar{M}_w$ are compared, their compliances should vary as $\bar{M}_z \bar{M}_{z+1}/\bar{M}_w^2$. Based on the success of the Rouse model mixing law for viscosity, one might hope for correlations of the form:

$$J_e^0 = J^*(\bar{M}_w) \frac{\bar{M}_z \bar{M}_{z+1}}{\bar{M}_w^2} \tag{5.29}$$

in which $J^*(M)$ is the compliance of a monodisperse polymer of molecular weight M, obtainable from the results of the preceding section. In some cases the polydispersity correction is fairly well represented by Eq.(5.29). For example, J_e^0 values in concentrated solutions of samples with exponential distribution ($\bar{M}_w/\bar{M}_n = 2$, $\bar{M}_z\bar{M}_{z+1}/\bar{M}_w^2 = 3$) are roughly three times the values for narrow distribution samples with the same $\bar{M}_w$ (178, 195, 211). Values for blends at low concentrations (198) are also in reasonable accord. However, there are indications that J_e^0 rises more rapidly than the Rouse factor $\bar{M}_z\bar{M}_{z+1}/\bar{M}_w^2$ for slight amounts of polydispersity (165). The Rouse factor again underestimates the polydispersity effect in binary blends of polystyrenes which are not too different in molecular weight ($M_2/M_1 \approx 2$), but gives reasonable results for poly(dimethylsiloxane) blends with $M_2/M_1 \approx 10$ (212). On the other hand, J_e^0 values for blends of polyisobutylene with $M_2/M_1 \approx 8$ are higher than predicted (213), while those for polyvinyl acetate blends agree fairly well with the Rouse factor over a wide range of M_2/M_1 and molecular weight levels (148).

Forms similar to Eq.(5.29) but involving different averages have been suggested. Mills (204) has proposed the form:

$$J_e^0 = J^*(\bar{M}_w)\left(\frac{\bar{M}_z}{\bar{M}_w}\right)^{3.7} \tag{5.30}$$

based on data for high molecular weight polyethylene, polystyrene, and polydimethylsiloxane. Zosel (194) proposed the same form but with an exponent 2.5 rather than 3.7 on the polydispersity factor. Leaderman et al. (213) correlated J_e^0/J^* with $\bar{M}_w/\bar{M}_n$, but the implied dependence on the properties of the low molecular weight tail (through $\bar{M}_n$) must be incorrect in general.

Phenomenological blending relations for J_e^0 have been suggested, based on the properties of η_0 and J_e^0 for narrow distribution systems and the assumption that η_0 always obeys Eq. (5.28) in blends. The relaxation spectrum for a binary system according to the linear mixing rule is (214)

$$H_b(\tau) = \varphi_1 H_1(\tau/\lambda_1) + \varphi_2 H_2(\tau/\lambda_2) \tag{5.31}$$

in which φ_1 and φ_2 are the volume fractions of the two components in the blend, $H_1(\tau)$ and $H_2(\tau)$ are their relaxation spectra before blending, and λ_1 and λ_2 represent the scaling factors by which their relaxation times are altered in the blend. The factors λ_1 and λ_2 are chosen to make the resulting expression for η_0 agree with Eq. (5.28). If the pure component spectra are chosen to give $\eta_0 \propto M^{3.4}$ and $J_e^0 \propto M$ (the shifted Rouse forms), then a polydispersity factor of $\bar{M}_z\bar{M}_{z+1}/\bar{M}_w^2$ follows directly. If $\eta_0 \propto M^{3.4}$ and $J_e^0 \propto M^0$ for the pure components (the forms actually observed for $M > M_c'$), the form

$$J_e^0 = J^*(\bar{M}_w)\left(\frac{\bar{M}_z}{\bar{M}_w}\right) \tag{5.32}$$

is calculated, which is too weakly dependent on polydispersity to be correct.

Higher order blending laws have been proposed (215). The quadratic law,

$$H_b(\tau) = \varphi_1^2 H_{11}(\tau/\lambda_{11}) + 2\varphi_1\varphi_2 H_{12}(\tau/\lambda_{12}) + \varphi_2^2 H_{22}(\tau/\lambda_{22}) \tag{5.33}$$

with the added stipulation that the cross term has the same form as the pure component terms, yields

$$J_e^0 = J^*(\bar{M}_w)\left(\frac{M_z}{M_w}\right)^2. \tag{5.34}$$

This polydispersity dependence is probably still too weak, perhaps because of the unjustified assumption about the form of H_{12}. The quadratic law leads directly to the experimentally observed behavior $J_e^0 \propto c^{-2}$ if one of the components is solvent; the linear law gives $J_e^0 \propto c^{-1}$.

There is still a great deal of disagreement about the systematics of poly-dispersity effects on J_e^0. The effects themselves are clearly large in concentrated systems, and the quadratic blending law appears promising if a satisfactory form for the interaction spectrum H_{12} can be found. Generalization of the quadratic law to arbitrary distributions, perhaps with the help of some approximate theory for H_{12}, would help greatly as a framework for systematizing the experimental observations.

6. Molecular Entanglement Theories of Linear Viscoelastic Behavior

The discussion on linear viscoelasticity of narrow distribution polymers has centered mainly on three properties, the plateau modulus G_N^0, the steady state viscosity η_0 and the steady state compliance J_e^0. Each property is associated with a characteristic molecular weight: M_e, M_c, and M_c' respectively in the undiluted state. Near M_c the viscosity—molecular weight relation changes from $\eta_0 \propto M$ to $\eta_0 \propto M^{3.4}$. For $M \gg M_c$ the relaxation time distribution is separated into two parts, slow relaxations which are highly sensitive to molecular weight and rapid relaxations which are insensitive to molecular weight. The modulus contributed by the slow relaxations is G_N^0; G_N^0 is independent of molecular weight for $M \gg M_c$, and $M_e \equiv \varrho RT/G_N^0$. Near M_c', the steady state compliance changes from direct proportionality to independence of molecular weight. The characteristic molecular weights stand in the order $M_e < M_c < M_c'$, are practically independent of temperature, are similar small multiples of each other for different polymers, and (in the case of M_e for example) correspond to roughly the same chain length from one polymer to another. Their values in concentrated solutions are accurately proportional to the reciprocal of polymer concentration and independent of solvent nature.

These properties, together with the order-of-magnitude agreement between viscosities for undiluted polymers below M_c and calculated values from the Rouse theory, suggest that behavior at high molecular weights is modified by an inter-action between the chains. This interaction is a function of the product cM for concentrated solutions and melts, and appears to be an intrinsic property of concentrated long-chain systems. A rope-like entangling of chains satisfies these qualitative requirements. The problem, of course, is to convert this idea into an acceptable theory. From the standpoint of linear viscoelasticity this means relating the relaxation time distribution in the terminal region to molecular weight and concentration. Before proceeding to mechanistic explanations, it is useful to consider what kinds of information are available on the terminal distribution without appeal to specific molecular arguments.

First, for $M \gg M_c$ the terminal relaxations are fairly well separated from the transition relaxations. In view of Eqs. (3.24) and (3.25), the values of η_0 and J_e^0 for $M \gg M_c$ are properties of the terminal relaxations alone. The mean relaxation times of the terminal distribution are therefore given by Eqs. (3.26)–(3.28), combined with the experimental results in Section 5.

$$\tau_n = \eta_0/G_N^0 = \frac{\eta_0 \, M_e \, \varrho}{c^2 \, R\,T},\tag{6.1}$$

$$\tau_w = \eta_0 \, J_e^0 = \frac{0.4 \, \eta_0 \, M_c' \, \varrho}{c^2 \, R\,T},\tag{6.2}$$

and

$$\frac{\tau_w}{\tau_n} = J_e^0 G_N^0 = 0.4 \, \frac{M_c'}{M_e} = 0.4 \, \frac{(M_c')_{\text{soln.}}}{(M_e)_{\text{soln.}}}.\tag{6.3}$$

Thus, what might be termed the relaxation time polydispersity, τ_w/τ_n, tends to a constant value for $M \gg M_c$ (see Table 5.5). The difference between M_e and M_c', [or $(M_e)_{\text{soln.}}$ and $(M_c')_{\text{soln.}}$] is a direct result of this polydispersity of relaxation times.

If the terminal spectrum were simply a shifted Rouse spectrum, τ_n would be the same as that given in Eq. (6.1) but τ_w and τ_w/τ_n would be different:

$$\tau_w = \frac{0.4 \, \eta_0 \, M}{c\,R\,T} \quad \text{(Shifted Rouse spectrum)},\tag{6.4}$$

$$\frac{\tau_w}{\tau_n} = \frac{0.4 \, c\,M}{\varrho \, M_e} \quad \text{(Shifted Rouse spectrum)}\tag{6.5}$$

so the relaxation time polydispersity would increase with both concentration and molecular weight, contrary to observation.

Any satisfactory rheological theory of entanglement for monodisperse chains must not only account for the observed shift in the mean relaxation time τ_n with concentration and molecular weight [Eq. (6.1)] but also for the relatively narrow distribution of relaxation times in the terminal spectrum. It must also conform in a natural way to the very sensitive relationship observed between relaxation time polydispersity and molecular weight distribution. There are also questions from the viewpoint of chemical structure. What is it that controls the value of M_e for a given polymer, and what precisely constitutes a chain entanglement? It is the kind of localized, rope-like interaction that comes to mind from the analogy with networks, or is it a more diffuse topological interaction that simply manifests itself rheologically as network-like response at short times and enhanced friction in steady flow?

This section covers the present status for both the rheological and structural questions. The first section reviews the progress on formal molecular theories which deal with the rheology of concentrated polymer systems without explicit consideration of the mechanistics of intermolecular interaction. Successful and truly predictive theories of that kind would of course remove much of the incentive for a detailed mechanistic discussion of entanglement. At the very least, they would serve to isolate particular characteristics of the interaction for examination. Following this is a discussion of network theories which take explicit account of interchain coupling by treating polymer liquids as relaxing network systems. This is followed by a review of several theories which seek to explain the shift in τ_n, actually the $M^{3.4}$ dependence of η_0 and the presence of a plateau modulus which is independent of M, through an enhanced frictional resistance to large scale motion. Next is a review of theories on the terminal relaxation spectrum, with particular attention to their consistency with the observed behavior of J_e^0. Following this is a comparison among theories which deal with the characteristic molecular weights. The last section is a brief review of theories dealing with the effects of polydispersity.

6.1. Formal Molecular Theories

Chikahisa (216) and Williams (217–219) have examined flow behavior in concentrated polymer systems without detailed consideration of the mechanism of intermolecular interaction. Williams explicitly limits his discussion to unentangled systems; Chikahisa uses an entanglement terminology, although not in a specific way. Both approaches grow out of the formalism which was developed to deal with transport properties in small-molecule liquids.

Williams begins with Fixman's equation (220) for the stress contributed by intermolecular forces in flexible chain systems. The theory assumes that the polymer concentration is high enough that intermolecular interactions control the stress. The shear stress contributed by polymer molecules in steady shear flow is expressed in the form

$$\sigma = \frac{1}{4} v^2 \iiint \left[x_1 \frac{\partial V}{\partial x_2} + x_2 \frac{\partial V}{\partial x_1} \right] g(\dot{\gamma}, x_i)\, dx_1\, dx_2\, dx_3 . \qquad (6.6)$$

The intermolecular potential $V(x_i, \dot{\gamma})$ is the energy associated with the interaction of a pair of molecules whose centers are separated by vector distance r with components x_1, x_2, x_3:

$$V = A \int f(r') f(r' + r) dr',$$ (6.7)

$f(r')$ being the segment density distribution function of the molecules. In the context of Fixman's theory, the potential has a thermodynamic origin. The parameter A governs the magnitude of the pair interaction energy at equilibrium:

$$A = M^2 kT \frac{d^2 \varepsilon^*}{dc^2}$$ (6.8)

and $\varepsilon^* kT$ is the free energy of mixing per polymer segment. The potential also depends on shear rate through the effect of the latter on the segment density distribution.

The pair correlation function g describes the distribution of molecular centers in the solution. In concentrated systems at rest, $g \approx 1$. Flow alters g, and it is this change which gives rise to the drag forces. For sufficiently slow shearing flows,

$$g(\dot{\gamma}, r) = 1 + C \frac{\zeta \dot{\gamma}}{kT} \frac{x_1 x_2}{(x_1^2 + x_2^2 + x_3^2)^{5/2}}$$ (6.9)

in which ζ is an intermolecular frictional coefficient and C is an undetermined parameter with dimensions of (length)^5.

Viscosity at low shear rates is obtained by substituting the results of Eqs. (6.8) and (6.9) into Eq. (6.6), and evaluating V from the segment density distribution of undisturbed random coils.

$$\sigma/\dot{\gamma} = \eta_0 = 2\pi c^2 \zeta C \frac{d^2 \varepsilon^*}{dc^2} \left(\frac{3}{4\pi S^2} \right)^{3/2} .$$ (6.10)

Williams then estimates ζ with Kirkwood's equation, employing the Fixman-Peterson expression for $g(r)$ in chain molecules. He arrives at a variation with molecular weight of the form $\zeta \propto M^{1/4}$. Williams then estimates the chain length dependence of C by a dimensional argument, concluding that $C \propto S^5 \propto M^{5/2}$. The final result is of the form

$$\eta_0 \propto M^{5/4} c^2 = (cM^{5/8})^2 .$$ (6.11)

Chikahisa (216) proceeds through a direct analogy with small molecule transport theory. At an intermediate stage of calculation, he arrives at an equation which is formally similar to Eq. (6.6):

$$\sigma = \dot{\gamma} \int P_1(r)\, P_2(r)\, dr \tag{6.12}$$

in which P_1 is the radial distribution function for molecular centers, and P_2 is a pair-wise intermolecular force. The force is taken to be purely frictional in nature, directly proportional to chain length, and proportional as well to the mean shared volume of the pair. Using dimensional arguments and an approximate expression for the radial distribution function, Chikahisa calculates an expression for viscosity in highly concentrated systems. In the terms used in this review, the result for $E = M/M_e - 1 \gg 1$ is:

$$\eta_0 = \frac{(6\pi)^{3/2}}{270}\, \zeta_0\, v^2\, S^5\, E^{3/2}\, n \tag{6.13}$$

which can be rearranged to give:

$$\eta_0 = \left(\frac{\zeta_0\, v\, S^2\, n}{6} \right) \left(\frac{(6\pi)^{3/2}\, v\, S^3\, E^{3/2}}{45} \right) \tag{6.14}$$

(Rouse term)   (Entanglement factor)

or $\eta_0 \propto M^3$ in undiluted systems.

It appears that the formal theories are not sufficiently sensitive to structure to be of much help in dealing with linear viscoelastic response: Williams' analysis is the most complete theory available, and yet even here a dimensional analysis is required to find a form for the pair correlation function. Moreover, molecular weight dependence in the resulting viscosity expression [Eq. (6.11)] is much too weak to represent behavior even at moderate concentrations. Williams suggests that the combination of variables in Eq. (6.11) may furnish theoretical support correlations of the form $\eta_0 = f(c[\eta])$ at moderate concentrations (cf. Section 5). However the weakness of the predicted dependence compared to experiment and the somewhat arbitrary nature of the dimensional analysis makes the suggestion rather questionable.

Unlike Williams, Chikahisa attempts to deal directly with entangling systems. The result is a viscosity expression which is not far from that observed experimentally, but the form is unfortunately more dependent on a series of intermediate assumptions about the nature of the friction forces than on the basic transport theory itself. Although not implausible, the assumptions are nevertheless arbitrary and lacking in theoretical justification.

6.2. Network Theories

Network theories view a highly entangled polymer system as a network structure with interconnections extending throughout the system. The entanglement junctions are the cross-links; the sections between junctions are the network strands. At equilibrium the concentrations of strands and junctions are constant, although individual elements of the network are transitory. Each is created by the thermal motions of the molecules, and each is eventually lost by the same process. Network theories employ the kinetic theory of rubber elasticity to calculate the mechanical properties in flow. The ends of each strand are assumed to move with the medium from the time the strand is created until it becomes inactive through the loss of one of its junctions. The stress is calculated as the sum of independent contributions from network strands of all ages.

The relaxing Gaussian network of Green and Tobolsky (4) is the earliest version of this model. Lodge (12) and Yamamoto (13) independently derived constitutive equations for similar systems, based on a stress-free state for each newly created strand and a distribution of junction lifetimes which is independent of flow history. For Gaussian strands in an incompressible system:

$$p_{ij} = -P_0\,\delta_{ij} + \int_{-\infty}^{t} M(t-t')\left[C_{ij}^{-1}(t',t) - \delta_{ij}\right]dt'. \tag{6.15}$$

The pressure P_0 represents the arbitrary additive contribution to the normal components of stress in an incompressible system, δ_{ij} is the Kronecker delta, $C_{ij}^{-1}(t',t)$ is the inverse of the Cauchy-Green strain tensor for the configuration of material at t' with respect to the configuration at the current time t [a description of the motion (221)], and $M(t)$ is the junction age distribution or memory function of the fluid.

Depending on the form chosen for the memory function, the Gaussian network model can show many of the qualitative features of polymer flow behavior (78): linear viscoelastic behavior for sufficiently small or sufficiently slow deformations, rubbery solid-like behavior for rapid strains, viscous liquid behavior for slow strain rates, and a positive first normal stress difference in steady shearing flow. On the other hand, the model yields a shear viscosity which is independent of shear rate, contrary to observed behavior. This defect can be remedied by relaxing the restriction to Gaussian chains or the assumption that the memory function is independent of flow history (222). In either case there is a substantial increase in complexity of the constitutive equation. Flory has criticized such theories (223) as being inconsistent with the rigorous theory of composite Gaussian networks. According to the latter, new strands which are created by cross-linking an already deformed network are not stress-free, as judged by the retarded stress-relaxation of a deformed system in which the junction concentration remains constant but individual junctions are lost and new ones formed at random. Generalization of the elastic liquid model to remove this inconsistency does not appear to have been made.

For our present purposes, the network theories suffer from an additional defect. They supply no information on the form of the memory function. The memory function must be obtained for each system by rheological experiments, and there is no way at present to predict how it should vary with the molecular structure of the polymer. For example, $M(t)$ can be obtained from the stress relaxation modulus $G(t)$:

$$M(t) = - \frac{dG}{dt} \tag{6.16}$$

and using Eq. (4.10), for example:

$$M(t) = \frac{1}{3} kT \sum_{i=1}^{N_T} e^{-t/\tau_i}/\tau_i . \tag{6.17}$$

If $M(t)$ is known, the behavior of the model in any flow situation can be calculated.

In network models the molecular arguments supply a form for the constitutive equation, but do not provide the detailed connections to molecular structure. As such, they provide a bridge between molecular theories which incorporate specific structural information in rather specific flow situations and continuum models which can generalize such information to arbitrary flows.

6.3. Entanglement Friction

Mutual obstructions to motion, such as would arise from a simple looping of chains, must be present in any system of interpenetrating chain molecules. In systems without permanent crosslinks these obstructions will only affect the number of pathways available for configuration change and can not in themselves affect the equilibrium distribution of configurations. It is important to distinguish between two properties of mutually excluded volume of polymer chains. One is the net excluded volume, a thermodynamic property which depends on the mixing characteristics of the components and affects equilibrium properties such as chain dimensions. The other is a physical exclusion of volume which *always* prevents simultaneous occupancy and thus precludes the passage of one chain directly through the backbone contour of another. Although the thermodynamic excluded volume is probably small in concentrated systems, physical excluded volume is always present, and chain connectedness will reduce the number of topologically acceptable pathways between large-scale configurations, thereby reducing the rates of configurational change.

If a steady systematic motion is imposed, those topological arrangements which present barriers to configuration relaxation will come into play. As in dilute solutions the chains are continually drawn into a disturbed distribution of configurations by the external motions and respond by diffusing collectively toward an equilibrium distribution. In concentrated systems each chain must select paths which carry it around the topological impediments presented by its surroundings. The result is a retardation of the collective response, appearing macroscopically as a frictional resistance to steady motion. The effect must of course be quite different in detail than simple Stokes friction for chains in a continuous Newtonian medium. Theories of entanglement friction are based on the hope [and some experimental evidence (33)] that these differences will cancel in the large, at least for sufficiently slow or sufficiently small deformations, and that entanglement effects can be represented by an enhancement in the molecular frictional coefficient. All such theories ignore random thermal motion and adopt a kind of "limp rope" picture for estimation of the topological contribution.

6.3.1. Theories of Entanglement Friction

The combined viscosity and diffusion measurements on concentrated systems by Bueche *et al.* (33), were described in Section 2. The results suggest that viscosity behavior above M_c can be explained simply in terms of an enhanced Stokes friction for the molecules. Bueche (7, 224) suggested that since polymer molecules in concentrated systems are looped through the coils of neighboring molecules, relative motion must be resisted by the need for these loops to slip around one another. Each molecule is assumed to participate in an average of E such temporary couples, creating (for $E \gg 1$) a temporary network of Gaussian segments joining $\dfrac{\nu E}{2}$ couples per unit volume. The coupling frequency is taken to be proportional to molecular weight. For $E \gg 1$ the molecular weight between coupling joints is $M_e = M/E$. In concentrated solutions, $E \propto cM$, assuming entanglement coupling to be simply a special kind of segment-segment contact (Section 5).

A systematic motion imposed on the center of gravity of any molecule in such a system induces additional systematic motions, that is, motions beyond those which would occur if the chain were being pulled through a monomeric liquid. Bueche separated these motions into two types, an induced motion in the surroundings caused by the tendency of the molecule to drag along its E coupling partner molecules, and a snaking circulatory motion of the molecule itself as it moves around the E coupling points. (When the molecules move simultaneously, as in shearing flows, each molecule in the system plays both roles and thus presumably partakes in both types of additional motion.)

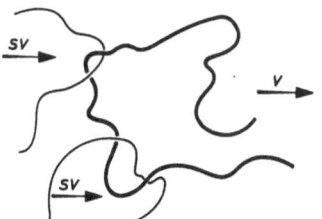

Induced Medium Motions

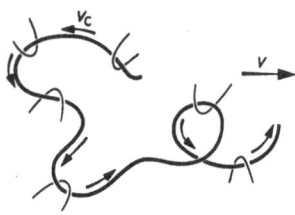

Circulatory Motions

The induced motions in the surroundings are estimated by examining the effects of a systematic velocity v imposed on one of the molecules, called here the central molecule. The couples of the central molecule lead to N_1 other molecules which move in response to the motion of the central molecule. Likewise, the N_1 first order partners are coupled to N_2 second order partners, and so on through a hierarchy of orders: order j contains N_j molecules, coupled for the first time to the central molecule through a sequence of j couples. The induced velocity of each order is assumed to be a constant fraction s of the velocity of the preceding order. The formal expression for the force resisting the motion of the central molecule becomes therefore:

$$F = \zeta_0 \, n \left[1 + \sum_{j=1}^{\infty} N_j \, s^j \right] v \tag{6.18}$$

in which n is the number of main chain atoms and ζ_0 is the frictional coefficient per main chain atom. The slip factor s $(0 < s < 1)$ is left as an undetermined parameter. [An average value $s = 4/9$ is found for pairs of looped ropes sliding past one another without contact friction in a viscous medium; s rises to $(4/9)^{1/2}$ if viscous drag is proportional to n^2 rather than n.]

Redundancy (multiple coupling between molecules or between members of different orders, coupling between molecules of the same order, etc.) reduces N_j below its maximum possible value, $E(E-1)^{j-1}$. Bueche asserts that the number of effective j-order couples is given by:

$$N_j = \int_0^{\infty} 4\pi \, r^2 \, \varrho_j(r) \, e^{-\varrho_j \,(r)/vE} \, dr \, . \tag{6.19}$$

The product vE is the total number of coupling units per unit volume in the system and $\varrho_j(r)$ is the concentration of potential j-order coupling sites at a distance r from the central molecule:

$$\varrho_j(r) = E(E-1)^{j-1} \left[\frac{3}{2\pi \, S^2 (2j-1)} \right]^{3/2} \exp\left[- \frac{3r^2}{2 \, S^2 (2j-1)} \right] . \tag{6.20}$$

Bueche's result (for $E \gg 1$, and with some harmless approximations to simplify integration) is:

$$N_j = \left(\frac{3}{2} \right)^{1/2} v E S^3 (2j-1)^{3/2} \, . \tag{6.21}$$

Insertion in Eq. (6.18) leads to (again for $E \gg 1$):

$$F = [K(s) \, v E S^3] \, n \zeta_0 \, v \tag{6.22}$$

in which

$$K(s) = \left(\frac{3}{2}\right)^{1/2} \sum_{j=1}^{\infty} s^j (2^j - 1)^{3/2}.$$ (6.23)

For the circulation contribution, Bueche treats the coupling points as fixed points around which the central molecule must move in order to proceed in the direction of motion. The segments between successive coupling sites are labeled $1, 2, ..., E/2$ out from the center of the molecule. Bueche argues that the speed of segment i relative to the medium, v_i, compared to v, the speed of the center of gravity, is given by:

$$v_i = i^{1/2} v.$$ (6.24)

The contribution of circulation to the frictional coefficient is then obtained by equating the rate of energy dissipation for the chain to the sum of the rates for the individual segments:

$$F \cdot v = 2 \sum_{i=1}^{E/2} n_e v_i^2 \zeta_0 = \frac{2\zeta_0 n}{E} v^2 \sum_{i=1}^{E/2} i$$ (6.25)

giving finally, for $E \gg 1$:

$$F = \left[\frac{E}{4}\right] n\zeta_0 v.$$ (6.26)

The combination of enhancement terms for the molecular friction coefficient from Eqs. (6.22) and (6.26) with the Rouse equation for viscosity yields

$$\eta_0 = \left(\frac{nvS^2\zeta_0}{6}\right) \times (K(s) vES^3) \times \left(\frac{E}{4}\right)$$ (6.27)

| Rouse term | Contribution from induced medium motion | Contribution from induced central molecule motion. |

When the molecular weight and concentration dependences of the structural parameters S^2, E, n, and v are inserted ($S^2 \propto M$, $E \propto cM$, $n \propto M$, $v \propto c/M$), an expression for viscosity in highly entangled linear molecules is obtained.

$$\eta_0 \propto \zeta_0 n v^2 S^5 E^2 \zeta_0 \propto \zeta_0 c^4 M^{3.5}.$$ (6.28)

De Gennes (225) has recently examined the diffusion of chains in a medium containing fixed, loop-like obstacles. In effect, the chain is confined within a tunnel defined by the obstacles along its contour. It moves by an inching displacement, called reptation, which proceeds by the diffusion of length defects or loops along the chain. Only the end of the chain has a choice of direction for subsequent movement as it moves out of the tunnel. The contour tunnel is therefore created continuously as the chain moves among the obstacles. The molecular friction coefficient of chains in this environment is proportional to $\zeta_0 n^2$, rather than $\zeta_0 n$ as in a viscous medium without obstacles. This increased power of n corresponds to a circulation effect since only the central molecule moves. As in the Bueche result, the circulation contribution is directly proportional to chain length.

The De Gennes method can be applied directly to Bueche's model without the defect diffusion formalism. Consider a chain with n main chain atoms, mean square end-to-end distance $\langle r^2 \rangle$, and a large number of entanglement obstacles along its contour. For simplicity, suppose the number of main chain atoms between successive obstacles and the scalar distance between obstacles are constants, $n_e = n/(E+1) \approx n/E$ and $r_e = \left(\dfrac{\langle r^2 \rangle}{E} \right)^{1/2}$ respectively. The obstacles are labelled $(i = 1, 2, 3, \ldots E)$ beginning at one end of the chain. The sections of chain between successive obstacles are also labeled; the vector distance from the i to the $i+1$ obstacle is a_i and the unit vector in that direction is a_i/r_e.

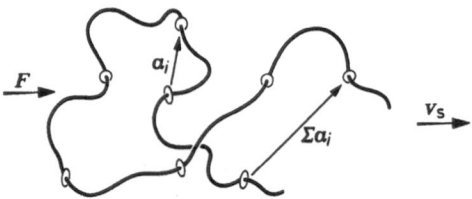

A force $F = F_s u$ is imposed on the chain (F/E on each segment between obstacles), F_s being the magnitude of the force and u a unit vector in the direction of the force. For any segment i, only the component of this force along direction a_i will be effective in moving the chain through its tunnel. The net force acting on the chain and moving it with a speed v through its tunnel is therefore:

$$\frac{F_s}{r_e E} \sum_{i=1}^{E} u \cdot a_i . \tag{6.29}$$

This force must be equal to the total frictional force for movement of the chain through the tunnel, $\zeta_0 n v$. Thus:

$$v = \frac{F_s}{\zeta_0 n r_e E} u \cdot \sum_{i=1}^{E} a_i = \frac{F_s r_s}{\zeta_0 n r_e E} \tag{6.30}$$

in which $r_s = \boldsymbol{u} \cdot \Sigma \boldsymbol{a}_i$ is the component of the end-to-end vector of the chain in the direction of the force. The component of the velocity in the same direction for the segment i is:

$$v_i = v \frac{a_i}{r_e} \cdot \frac{\Sigma a_j}{r_s} . \tag{6.31}$$

The motion of the center of gravity of the chain must always be in the direction of its end-to-end vector $\Sigma \boldsymbol{a}_i$. The velocity component in the direction of the force is therefore:

$$v_s = \frac{\Sigma v_i}{E} = \frac{F_s r_s^2}{\zeta_0 n r_e^2 E^2} . \tag{6.32}$$

The average over all chains in the system (presenting all possible directions of end-to-end vectors) yield $\langle v_s \rangle$. With $\langle r_s^2 \rangle = \langle r^2 \rangle / 3 = E r_e^2 / 3$, Eq. (6.32) rearranges after averaging to give

$$F_s = 3 n E \zeta_0 \langle v_s \rangle \tag{6.33}$$

which provides a circulation factor of $3E$, the same form as Bueche's expression but with a different numerical coefficient.

Eyring *et al.* (226) examine the entanglement problem from a somewhat different point of view, the activated complex theory of liquid viscosity. In a monomeric liquid the molecules move by random jumps from one equilibrium position to another. The jump frequency is controlled by an activation barrier between neighboring sites. According to activated complex theory, a shear stress lowers the barrier in the direction of the stress and raises it in the opposite direction, producing a bias in jump frequency and a net flow of molecules in the stress direction. For low stresses, the expression for viscosity in a monomeric system is:

$$\eta_m = \frac{6kT}{K_0} \frac{\lambda_2}{\lambda_1 \lambda_3 \lambda^2} \tag{6.34}$$

in which K_0 is the equilibrium jump frequency, λ is the jump distance and λ_1, λ_2, and λ_3 are the distances separating layers of molecules in the three coordinate directions.

If the molecules are now joined to form chains n units long, the jump frequency per chain is $K_0 n$, while the jump distance (displacement in the direction of the jump) of the center of gravity of the chain is λ/n. The zero shear viscosity of a collection of independent chains, comprised of units which themselves move independently, is therefore:

$$\eta_0 = \eta_m n.$$

(6.35)

The viscosity is thus directly proportional to chain length, as in the Rouse model. The authors argue however in favor of an additional factor proportional to $n^{1/3}$ because the distance between molecular centers in the direction of the velocity gradient, λ_2, increases by this factor as n increases. Their result is therefore

$$\eta_0 \propto n^{4/3} \propto M^{4/3} = M^{1.33}$$

(6.36)

for non-interacting chains.

Eyring and co-workers extend the calculation to entangled systems, postulating that E coupling points are distributed along the length of each chain. These junctions divide the chain into $E+1$ chain sections, each section containing n_e units. They argue that the average displacement of the center of gravity per jump is $\lambda/n(E+1)$ rather than λ/n because only one of the $E+1$ chain sections contains the center of the molecule. Their expression for viscosity becomes

$$\eta_0 = \eta_m n E^2$$

(6.37)

or, employing the form which contains the extra $n^{1/3}$ factor,

$$\eta_0 \propto n^{4/3} E^2 \propto M^{10/3} = M^{3.33}$$

(6.38)

for undiluted polymers.

Thus, in effect, Eyring and co-workers ascribe all effects due to entanglement to additional motion of the central molecule, obtaining thereby a circulation term proportional to E^2. Graessley (227) on the other hand lumps all extra motion in the molecules of the medium. Again, each central molecule is assumed to participate in E coupling interactions. Each couple involves a different molecule moving through the environs of the central molecule in simple shearing flow. Each passing molecule is divided into two parts by its couple with the central molecule: a longer section L containing n_L main chain atoms, and a shorter section S containing n_S atoms ($n_S + n_L = n$). Likewise, there are E_S entanglement junctions along S and E_L along L, so $E_S + E_L = E - 1$. Each passing chain undergoes additional motion as it slips around its junction with the central molecule. This extra motion is assumed to be furnished mostly by the short segment S, which as a consequence moves with speed v_S relative to the medium.

If segment S were coupled to the central molecule only, its segments would simply drift as a particle cloud towards the junction and the drag force on the central chain would be approximately

$$F = 2\zeta_0 n_S v_S. \tag{6.39}$$

Because S is coupled with other molecules, its segments must traverse a circulating path in moving towards its junction with the central molecule. With the circulation contribution included, the drag force on the central molecule becomes

$$F = 2\zeta_0 n_S E_S v_S. \tag{6.40}$$

Finally, v_S is itself a circulation velocity with respect to the motion of section L, which is assumed to proceed steadily with the velocity of the medium at the center of gravity of the passing chain v:

$$F = 2\zeta_0 n_S E_S E_L^{1/2} v. \tag{6.41}$$

The average for all possible values of $n_S(0 < n_S < n/2)$, with $E_S = n_S/n_e$ and $E_L = n_L/n_e = (n - n_S)/n_e$, becomes finally

$$\zeta_e = 4\left[\frac{16}{105} - \frac{71}{2^{1/2}420}\right] n\, E^{3/2} \zeta_0 \tag{6.42}$$

or

$$\zeta_e = 0.131\, n\, E^{3/2} \zeta_0. \tag{6.43}$$

The quantity ζ_e is the mean frictional coefficient contributed by each entanglement junction, defined in terms of relative chain velocities.

Spring-bead models relate frictional force to the relative velocity of the medium at the point of interaction. The entanglement friction coefficient above is defined in terms of the relative velocity of the passing chain. Since the coupling point lies, on the average, midway between the centers of the two molecules involved, the macroscopic shear rate must be doubled when applying the result to a spring-bead model. Substitution of $2\zeta_e E$ for $\zeta_0 n$ in the Rouse expression for viscosity yields

$$\eta_0 = 0.262\, \frac{n\zeta_0 S^2 v}{6} E^{5/2}. \tag{6.44}$$

The usual concentration and molecular weight dependence of the structural variables thus combine to give

$$\eta_0 \propto \zeta_0 c^{3.5} M^{3.5}. \tag{6.45}$$

Aharoni has recently proposed a theory of flow based on entanglements between loops on the surface of collapsed coils (*43, 228*). The basic picture of amorphous polymer structure is almost certainly incorrect (see Section 2), and the derivation of viscosity is even more speculative than the others in this section.

6.3.2. Discussion of the Friction Theories

All the above theories attribute the enhanced friction in entangled systems to extra motions. In a crude way they speak to the main issue, which is the loss of relaxation pathways due to connectedness in the chain environment. However, all are rather arbitrary: they select certain types of motion for examination and exclude others, while in fact the motions must be cooperative and interdependent. In addition, the theories have features which appear to be incorrect, or at least inconsistent, even within the limited realm of motions examined.

Bueche's analysis of medium motion uses Eqs. (6.19) and (6.20) to evaluate N_j, the number of j generation chains coupled to the central chain. Equation (6.19) is itself rather arbitrary, and, moreover, Eq. (6.20) ignores attrition of lower order couples due to multiple coupling. That is, $(E-1)N_{j-1}$ should replace $E(E-1)^j$ in Eq. (6.20) for the calculation of $\varrho_j(r)$, and the N_j population should be obtained sequentially from the N_{j-1} population. Furthermore, even if the general form of Bueche's weighting function [Eq. (6.19)] is accepted, the exponent should logically be ϱ_j/v, rather than ϱ_j/vE, since the probability of redundancy in coupling depends on the ratio (density of potential j-order couples)/(density of molecules) rather than (density of potential j-order couples)/(density of total couples). If this substitution is made, a factor E is lost in the final expression for η_0, giving $\eta_0 \propto M^{5/2}$ instead of $M^{7/2}$.

A more direct argument for $M^{5/2}$ dependence as the outcome of Bueche's theory can be made. The total number of coupling points on any central molecule is directly proportional to molecular weight M. The number of polymer molecules lying within its pervaded volume (potential first generation partners) is proportional to vS^3. For random coil molecules vS^3 is proportional to $M^{1/2}$. Thus, for sufficiently large molecular weight, the number of couples on the central molecule will exceed the number of molecules available for coupling. At this point the effective number of first generation couples must become proportional to $vS^3 \propto M^{1/2}$. The volume pervaded by potential j-order couples is proportional to $(2j-1)^{3/2}S^3$, so the number of molecules available for j-order couples is $vS^3[(2j-1)^{3/2} - (2j-3)^{3/2}]$. Thus, in the limit of sufficiently long chains,

$$N_j \propto vS^3[(2j-1)^{3/2} - (2j-3)^{3/2}]. \tag{6.46}$$

With Eq. (6.18),

$$F \propto \zeta_0 n K(s) vS^3 v. \tag{6.47}$$

A factor of E has thus been lost from the frictional coefficient [compare Eq. (6.22)] so the viscosity is only proportional to $M^{5/2}$.

The Eyring analysis presents some problems also. Suppose one accepts their suggestion that only the movement of segments between one pair of couples, that containing the chain center, serves to advance the chain. Then, to be consistent, the correct modification of the viscosity equation for small molecules is to replace K_0 by $n_e K_0$ and λ by λ/n, rather than the original substitutions, $K_0 \to n K_0$ and $\lambda \to \lambda/n(E+1)$. The result is

$$\eta_0 = \eta_m n \frac{n}{n_e} = \eta_m n E . \tag{6.48}$$

Thus, with this modified analysis the circulation contribution is proportional to E, in agreement with the Bueche, De Gennes and Graessley results. However, now a power of molecular weight is lost in the viscosity,

$$\eta_0 \propto M^2 \tag{6.49}$$

or with the $n^{1/3}$ term included[4],

$$\eta_0 \propto M^{7/3} = M^{2.33} \tag{6.50}$$

Graessley's analysis neglects both redundancy in coupling and circulation of the central molecule. The interaction in shearing flow takes place between an unyielding central molecule and coupled passing molecules. The latter circulate independently through their own couples with the medium, producing slip at the junction with the central molecule. This distinction between central chain and passing chain is somewhat artificial. Since all chains in the system are equivalent, the additional motions generated by such pair-wise interactions must be independent of wich chain is chosen to be central. A later analysis (212) removes

---

[4] The Eyring equation for the viscosity of long chains without entanglements turns out to have the same form as the Rouse equation if λ is taken to be the length of an equivalent bond in the random coil molecule, n the number of such equivalent bonds, and $\lambda_1 \lambda_2 \lambda_3 = \lambda^3$ the volume per chain unit comprising an equivalent bond. The self-diffusion coefficient D for a freely jointed chain of n bonds is $K_0 \lambda^2/6n$, and its radius of gyration S is $(n\lambda^2/6)^{1/2}$. In an undiluted system, the number of chains per unit volume v is $1/\lambda^3 n$. The combination of these relations with Eqs. (6.34) and (6.35) and the Einstein equation, $D = kT/\zeta_0 n$, gives

$$\eta_0 = 6n\zeta_0 S^2 v . \tag{6.51}$$

Aside from the numerical factor, Eq. (6.51) is the Rouse formula. The fact that the Rouse form works well for short chains at constant ζ_0 casts doubt on the need for an extra factor of $n^{1/3}$.

part of this deficiency by examining the couples themselves, requiring the additional motion to be assumed by whichever molecule of the pair contains the shortest strand leading from the junction. However, the response due to each entanglement is still calculated separately; the additional motions caused by other entanglements along the same chain are ignored.

Thus, all present theories of the molecular frictional coefficient have rather serious deficiencies. Physical and mathematical approximations are made throughout, and there is really no independent way to justify or evaluate their importance. Ziabicki and Takserman-Krozer (229–231) have pointed out a number of difficulties in bringing dilute solution concepts to bear on concentrated systems through lumping intermolecular effects into the medium by simple averaging. Although their conclusions about such approaches are perhaps too pessimistic, it is nevertheless true that the present theory of entanglement friction is far from satisfactory.

It is not clear how improvements can be made without real progress on the difficult fundamental problems of diffusion in media with obstacles and co-operation of large-scale motions between interpenetrating chains which do not violate chain connectivity. The De Gennes reptation model (225) makes a significant contribution to the first problem, although in an admittedly simplified system. Rigorous calculations or computer simulations on well-defined models which relate to the second problem would be extremely valuable, even if the models themselves were not completely faithful representations of the assumed physical situation. It is not obvious how even to pose solvable problems, simplified or not, which relate to interchain cooperation.

6.4. Theories of the Terminal Relaxation Spectrum

Bueche's theory attributes the enhanced viscosity of entangled systems to an increase in the molecular frictional coefficient. With the aid of the Rouse model, Ferry et al. (11) extended this idea to other viscoelastic properties. They suggested that entanglement friction influences only the large-scale motions of each chain, thereby dividing the relaxation times of the Rouse model into two groups. Relaxation times associated with the higher normal modes, corresponding to cooperative motions over distances smaller than the entanglement spacing, are assumed to be unaffected by entanglement. Lower mode motions, governing cooperation which extends beyond the entanglement spacing, are slowed and their relaxation times shifted to larger values.

When E is large ($M \gg M_c$) the relaxation times of the affected modes are multiplied by the friction factor ratio $\left(\dfrac{\eta_0}{\eta_c}\right)\dfrac{M_c}{M}$, η_c being the viscosity at M_c. This assumption accounts for many experimental observations. It automatically builds in the correct molecular weight dependence of the viscosity. The gap in the relaxation time distribution creates a plateau region, and the transition relaxations comprise simply an unshifted Rouse spectrum. The plateau widens with increasing molecular weight, and the modulus in the plateau region, being dependent only on the chain length between entanglements, is independent of molecular weight. However, the spacings of the terminal spectrum are incorrect. As noted earlier, the experimental spectrum is much narrower than the shifted Rouse spectrum which results from this procedure.

Chömpff and Duiser (232) analyzed the viscoelastic properties of an entanglement network somewhat similar to that envisioned by Ferry *et al.* Theirs is the only molecular theory which predicts a spectrum for the plateau as well as the transition and terminal regions. Earlier Duiser and Staverman (233) had examined a system of four identical Rouse chains, each fixed in space at one end and joined together at the other. They showed that the relaxation times of this system are the same as if two of the chains were fixed in space at both ends and the remaining two were joined to form a single chain with fixed ends of twice the original size.

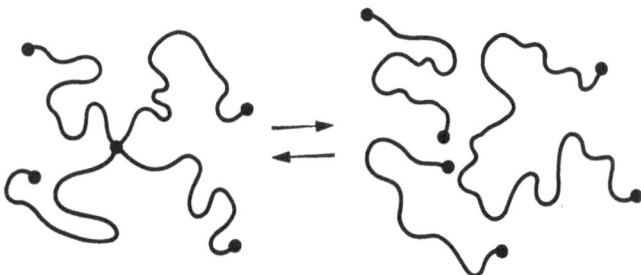

Chömpff and Duiser applied this tetrahedral model to entanglement networks. The suggest that the entanglement junction behaves as a site of large frictional resistance for the motion of one chain relative to the other. They then show that the relaxation times of an entanglement tetrahedron with equal strand lengths are the same as if one chain possessed a large frictional resistance for motion relative to the medium at the entanglement site and the other was un-

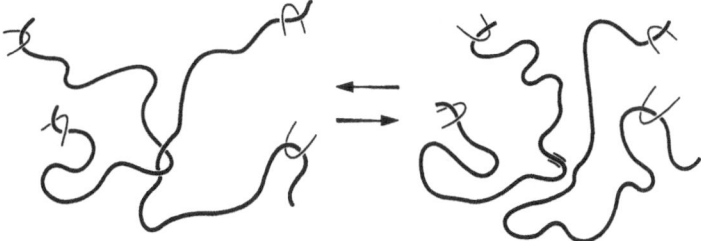

constrained. They use this result to transform a two dimensional network of strands between entanglement junctions into a mechanically equivalent system of independent primary chains with varying numbers of enhanced friction sites (slow points) along their lengths. The relaxation spectrum of the network is then formed from the spectra of independent chains in the equivalent system.

The network is decoupled by successively replacing junction tetrahedra by two independent strands, each terminated by remaining junctions and one with a slow point in the middle. The order of decoupling and assignment of slow points must be performed in a rather special way: the symmetry of remaining junctions must be preserved so that the same decoupling procedure can be applied to them in turn.

The transition and plateau spectra resulting from this procedure are, for $E \gg 1$, the same as those for a permanently crosslinked system with crosslink density equal to the entanglement density. The transition spectrum corresponds to a set of identical Rouse strands with fixed ends. In the continuous representation [Eq. (4.33)], $H(\tau) \propto (\tau_1/\tau)^{1/2}$, τ_1 being the longest relaxation time associated with fixed-end strands as long as those of the original network. The plateau spectrum arises mainly from cooperative motions extending beyond junction points in the network. It is formed from the slow relaxations of long strands between slow points in the equivalent system. Thus is depends directly on the concentration of strands in the equivalent system which contain multiples of the original network strands between slow points. For $E \gg 1$, Chömpff and Duiser derive the plateau form, $H(\tau) \propto \tau_1/\tau$, intersecting the transition spectrum near $\tau = \tau_1$, and extending from that point to longer times.

The terminal spectrum is furnished by cooperative motions which extend beyond slow points on chain in the equivalent system. The modulus associated with the terminal relaxations is $\frac{1}{2} vEkT$, which is smaller by a factor of two than the value from a shifted Rouse spectrum. It is consistent with a front factor $g = \frac{1}{2}$ given by some recent theories of rubber elasticity (Part 7). The terminal spectrum for $E \gg 1$ has the Rouse spacings for all practical purposes, shifted along the time axis by an undetermined multiplying factor (essentially the slow point friction coefficient). Thus, the model does not predict the terminal spectrum narrowing which is observed experimentally.

The Chömpff-Duiser procedure of symmetry-preserved decoupling does not appear to be applicable in an easy way to random networks in three dimensions. Junctions which anchor strands of unequal length can be decoupled, but with a considerable increase in difficulty (234). On the other hand, Chömpff and Prins (235) have introduced a degree of randomness in the decoupling procedure (the distribution of slow points within each uncoupled chain of the equivalent system was taken to be random) without altering the results appreciably. Random decoupling is much easier to implement, as shown below.

Consider the random decoupling of a system of v primary Rouse molecules, each containing E entanglement points, with random assignment of the slow points. In the equivalent system, the concentration of strands between successive slow points which are made up of exactly i of the original vE network strands is simply (for $E \gg 1$):

$$N_i = vE/2^{i+1} . \tag{6.52}$$

The transition and terminal spectra are then given by

$$H(\tau) = \sum_{i=1}^{\infty} H_i(\tau) \tag{6.53}$$

in which, using the continuous representation of the Rouse spectrum [Eq. (4.33)]:

$$H_i(\tau) = \frac{N_i k T}{2} (\tau_i/\tau)^{1/2} \qquad \tau < \tau_i/a^2$$

$$H_i(\tau) = 0 \qquad\qquad\qquad \tau > \tau_i/a^2 \qquad\qquad (6.54)$$

$$\tau_i = \tau_1 i^2$$

and τ_1 is the longest relaxation time of an origin network strand with fixed ends. Combination of Eqs. (6.52)–(6.54) and summation yields

$$H(\tau) = \frac{vEkT}{2} \left(\frac{\tau_1}{\tau}\right)^{1/2} \frac{I+1}{2^I} \qquad\qquad (6.55)$$

in which I is the first whole number whose magnitude exceeds a $(\tau/\tau_1)^{1/2}$. If I is replaced by $(\tau/\tau_1)^{1/2}$ to make an approximate continuous representation of the spectrum,

$$H(\tau) = \frac{vEkT}{2} \left(\frac{\tau_1}{\tau}\right)^{1/2} \left[1 + \left(\frac{\tau}{\tau_1}\right)^{1/2}\right] e^{-(\tau/\tau_1)^{1/2} \ln 2}. \qquad (6.56)$$

This spectrum departs somewhat from the Chömpff-Duiser result in the plateau region. The intensity remains larger in the vicinity of $\tau = \tau_1$, but then falls below the values for $\tau \gg \tau_1$. The terminal (shifted) spectrum for random decoupling is unfortunately still Rouse-like.

The reptation model (225) also appears to produce a Rouse spectrum at long times. In order to renew its configurations a chain must diffuse out of the tunnel defined by the fixed obstacles along its length. De Gennes calculates the auto-correlation function for the end-separation vector, obtaining

$$C(t) = \frac{8}{\pi^2} \sum_{\text{odd } i} \frac{1}{i^2} e^{-i^2 t/\tau_r} \qquad\qquad (6.57)$$

in which τ_r is the configurational renewal time. Equation (6.57) is the same form as $C(t)$ for the Rouse model [Eq. (4.7)], with $\tau_r = 2\tau_1$. The renewal time is a measure of the time required by a chain to diffuse out of its original tunnel. De Gennes shows that $\tau_r \propto M^3$, a result which can be justified physically by the following argument. To diffuse completely out of an original tunnel a chain must move the length of the tunnel L, which is proportional to the molecular weight of the chain. The diffusion coefficient D along this path is simply $kT/\zeta_0 n$. In order of magnitude one must have $D = L^2/\tau_r$, so

$$\tau_r \propto L^2/D \propto M^3 . \tag{6.58}$$

The longest unshifted Rouse relaxation time is only proportional to M^2. The extra power of M in the relaxation time is consistent with the circulation contribution to η_0 which was described in the preceeding section. This modification together with Eq.(6.57) suggests a simple Rouse spectrum shifted to longer times.

Edwards and Grant (*236, 237*) have very recently proposed a new calculation of self-diffusion and viscosity in highly entangled systems. They extend the reptation picture of De Gennes by allowing the surroundings (the tunnel) to diffuse as well as the chain in the tunnel. They suggest that in fact the rate limiting process may be the cooperative diffusion of tunnel and chain rather than chain reptation through the tunnel, and obtain $\eta_0 \propto M^3$ for this case. A number of intermediate assumptions are made in this calculation, and it is not easy to judge their physical significance. The terminal spectrum obtained is of the form

$$\tau_i = \tau_1/i^4 \qquad i = 1, 2, \ldots$$

in which τ_1 is the longest relaxation time. This spectrum has even wider spacings than a shifted Rouse spectrum, at variance with the narrower terminal spectrum demanded by the J_e^0 behavior.

Forsman and Grant (*238*) have recently proposed to handle the effects of entanglement by adding an extra force to the Rouse analysis. The force on the ith bead ($i = 0, 1, \ldots$) due to entanglement at the jth bead of a passing molecule is expressed as the product of an unspecified coupling coefficient N_{ij} and the frictional force acting at the jth bead. Utilizing the balance of forces on each bead, the frictional force on bead j is replaced by the sum of its osmotic and spring forces. The result on the spectrum is the same as that obtained by replacing the matrix A [Eq.(4.3)] of the Rouse theory by a new matrix $A - M$. The elements of M are composed of combinations of the coupling matrix N.

The symmetry properties of M alone are sufficient to prove that only the odd eigenvalues of A are affected by subtraction of M. Thus, according to this view, only the relaxation times of odd order are affected by entanglement coupling. Furthermore, the shifts diminish rapidly as mode number increases, so only the longest odd relaxation times are affected.

This result is intriguing because the unshifted long modes could account for the plateau relaxations. Also, the fact that only half the modes are shifted is reminiscent of the Chömpff-Duiser result, that the plateau modulus is only one half the value given by the conventional kinetic theory of elasticity. Unfortunately the coupling coefficients are unspecified, so the relative shifts and resulting spacing of the relaxations at long times are not given by the theory. An experimental test was performed (239) in which $G''(\omega)$ data on several concentrated polystyrene solutions were fitted to determine the relaxation time distribution. The results were consistent with shifts of the odd-order modes only, but other interpretations seem possible also. Unless the shifts are such as to narrow the spacings, the theoretical compliance J_e^0 will not match experimental behavior. If the same shift factor applies to all odd relaxations the reduced compliance approaches $J_e^0 cRT/M = \Sigma_{\mathrm{odd}\,i}\,i^{-4}/(\Sigma_{\mathrm{odd}\,i}\,i^{-2})^2 \approx 0.72$ for $M \gg M_c$. If the shift factor decreases with increasing mode number, the spectrum broadens.

Thirion (239a) has suggested that the plateau and terminal regions are the result of diffuse interchain interactions in a viscoelastic medium. He obtains a modified Rouse spectrum by replacing the subchain frictional coefficient by a time dependent "micro-memory" function. The theory is partly phenomenological since the memory function is not specified. However, reasonable choices lead to forms for $G'(\omega)$ and $G''(\omega)$ which are similar to those observed experimentally.

In view of the various suggestions on modifying the Rouse model to account for terminal zone spectrum in entangled systems, it is worth examining what effects can be produced by other modifications, particularly changes in the matrix which controls the terminal relaxation time spacings. The terminal relaxations of the Rouse model with E beads or slow points are given by the non-zero eigenvalues of the $E \times E$ matrix A [Eq. (4.3)]. Chömpff and Duiser and Ferry et al. obtained these same eigenvalues and therefore the shape of the terminal is unchanged. Forsman and Grand require the eigenvalues of a difference matrix, $A - M$. Depending on the choice of M, the spacing of the altered eigenvalues could be either greater or less than the Rouse spacings. Hoffman (240) and Chömpff and Prins (235) have speculated that the bead mobilities may depend on the position of the bead along the chain. According to Graessley (212), the mean entanglement frictional coefficients at the ends of the chain are small but they grow to large values near the chain center, rising as approximately the square of the contour distance from the chain end. The matrix governing the eigenvalues in this case is $B \cdot A$, in which B is a diagonal matrix containing the mobilities (reciprocal of the frictional coefficients) of the successive beads.

Some narrowing of the spectrum does occur when there is a strong positional dependence in the frictional coefficient. For example, the reduced compliance is:

$$J_{eR} = \frac{J_e^0 cRT\eta_0^2}{M(\eta_0 - \eta_s)^2} = \frac{\Sigma\,1/\lambda_i^4}{(\Sigma\,1/\lambda_i^2)^2} \tag{6.59}$$

where the λ_i are the non-zero eigenvalues of $B \cdot A$. For the usual Rouse model, the ζ_i are all equal and $J_{eR} = 0.4$. The narrowest possible distribution would result if all eigenvalues were the same, in which case $J_{eR} = 1/E$, which is the form observed experimentally. If ζ_i grows as the square of the distance from the nearest chain end, J_{eR} is 0.31 for large E; for a cubic dependence, the value is 0.30 (241). Vinogradov and co-workers (242) have reported some results by Pokrovsky in which the frictional coefficients increase as the 2.4 power of distance measured from the center of the chain. The reduced compliance is reported to be slightly smaller, 0.315 in this case, but independent of E for $E \gg 1$.

In view of the comparative intensitivity of J_{eR} to the properties of B, it seems unlikely that terminal spectra which are narrow enough to agree with experimental compliance data (that is, to make J_{eR} inversely proportional to E) can be produced merely by introducing a distribution of frictional coefficients within the molecule. The same holds true if the spring constants of the Rouse model are allowed to vary with position (241).

A terminal spectrum of about the right shape is obtained if each entanglement is treated as a separate bead-spring interaction between the molecule as a whole and the medium (212).

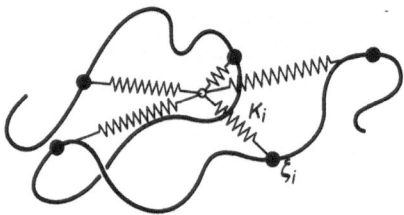

The beads represent entanglement sites which are distributed uniformly along the chain contour; the frictional coefficients increase rapidly with distance from the chain ends. The spring constant also depends on contour position, being governed by the mean equilibrium distance of that position from the center of gravity. The resulting spectrum is narrower than the Rouse spectrum, and for $E \gg 1$:

$$J_e^0 = \frac{1.798}{\nu E k T},\qquad(6.60)$$

or

$$J_e^0 \propto \frac{M}{E c R T} \propto \frac{1}{c^2 T}\qquad(6.61)$$

as observed experimentally. The long relaxation times are also of the experimental form:

$$\tau_w \propto \frac{\eta_0 M}{E c R T} \propto \frac{\eta_0}{c^2 T}.\qquad(6.62)$$

The failure of the Rouse theory was attributed to the pathological nature of medium motions in entangled systems, and not any special defect in the Rouse representation of the polymer chain itself. For Rouse chains in a deforming continuous medium, the frictional force depends on the systematic velocity of the bead relative to the medium. The frictional force on a bead is therefore a smoothly varying function of its internal coordinates. In entangled systems each interaction site is well removed from the molecular centers. of both chains involved. Thus, even if the molecular centers move affinely, the frictional forces at successive sites along a chain will change abruptly both in direction and magnitude. Unfortunately, attempts to build this effect into the analysis while retaining the Rouse model of the chain itself, for example by allowing the velocity gradient to fluctuate from position to position along the chain, have not been successful. The suggested loss of long-range correlations along the chain is therefore unconfirmed, and the model modification that leads to Eqs. (6.60)–(6.62) must be regarded as speculative.

Hayashi (243) has modified the Rouse model to account for both the plateau and the observed forms of τ_1 and J_e^0. Following Ferry, he takes the medium to be purely viscous for the high modes with a segmental frictional coefficient which is independent of molecular weight. For correlated motions which extend beyond the entanglement spacing he suggests that the medium is viscoelastic rather than purely viscous as in the usual Rouse formulation. He uses $\zeta_e \propto M^{2.5} \zeta_0$ and makes the effective spring constants of the normal modes increase with decreasing mode number. The frictional coefficient substitution is understandable, and amounts to the original Ferry modification. The justification for modifying the spring constants in the manner chosen is not clear. In any case, the modification should take place before the transformation to normal coordinates, so that the physics of the assumption can be seen. The result of the assumption is a narrowed relaxation spectrum, with $J_{eR} \propto 1/E$, as observed experimentally.

Hayashi later suggests (244) that E is proportional to c at moderate concentrations and to c^2 at high concentrations, concluding therefore that J_e^0 progresses from c^{-1} to c^{-2} and finally c^{-3} dependence as concentration approaches the undiluted state. The prediction of c^{-3} dependence seem to rest on the premise that the number of junctions per unit volume Ev becomes proportional to c^3 at high concentrations. The physical reasoning for any dependence other than c^2 is not made clear in the paper, and the result should probably be rejected, at least insofar as constituting a theoretical conclusion. As pointed out earlier, most data support $J_e^0 \propto c^{-2}$ and $E \propto c$ in highly concentrated systems.

Williams has derived the molecular weight and concentration dependence of a viscoelastic time constant τ_0 (actually the characteristic time governing the onset of shear rate dependence in the viscosity) from his theory (217–219). Employing a dimensional argument, he equates the parameters which control the shear rate dependence of chain configuration and the intermolecular correlation function. The result agrees with the observed form of characteristic relaxation time in concentrated systems [Eq. (6.62)]:

$$\tau_0 \propto \eta_0 / c^2 T. \tag{6.63}$$

The theory is not detailed enough to predict either numerical coefficients or the relaxation time distribution. Examination of the dimensional argument suggests that the form in Eq. (6.63) is probably not uniquely determined by the theory.

6.5. Theories of the Characteristic Molecular Weights

6.5.1. Relationships among M_e, M_c, and M_c'

Each of the viscoelastic parameters G_N^0, η_0, and J_e^0 has associated with it a characteristic molecular weight which either measures an equivalent spacing of entanglement couples along the chain (M_e, deduced from G_N^0 with the kinetic theory of rubber elasticity), or marks the onset of behavior attributed to the presence of entanglements (M_c and M_c', deduced from η_0 and J_e^0 as functions of molecular weight). Table 5.2 lists M_e, M_c, and M_c' for several polymers. Aside from certain difficulties in their evaluation, each is a rather direct and independent reflection of experimental fact.

The characteristic molecular weights seem clearly related to the same under-lying set of interactions, and their relative values provide useful tests for proposed theories. In particular, if the interactions are strictly non-specific and topological, then one might expect M_c/M_e and M_c'/M_c to be the same for all linear polymers. Differences among polymers would be reflected only in the differences of, for example, their individual values of M_e. The latter differences might in turn be expected to be related in some manner to the configurational and space-filling properties of the chains of different polymers. Berry and Fox (16) point out that M_c is approximately $2 M_e$ for a variety of non-polar polymers. In mildly polar polymers such as polydimethyl siloxane and polyvinyl acetate M_c appears to exceed $2 M_e$ to some extent. In the case of polymethyl methacrylate M_c/M_e may be as large as 5–8. Values of M_c' appear to be approximately 3–$4 M_c$, although in polyisoprene the relationship appears to be more nearly $M_c' = 6 M_c$.

Several of the entanglement friction theories predict values for M_c/M_e. The ratio is obtained by setting equal the Rouse equation for η_0 and the corresponding theoretical expression for highly entangled systems. This is equiv-alent to evaluating M_c experimentally from the intersection of the low and high molecular weight portions of η_0 vs. M graphs.

$$\text{Bueche (224):} \quad \frac{K(s)}{4} \varrho E_c^2 \left(\frac{S^2}{M} \right)^{3/2} M_c^{1/2} = 1$$

$$\text{Chikahisa (216):} \quad \frac{(6\pi)^{3/2}}{45} \varrho E_c^{3/2} \left(\frac{S^2}{M} \right)^{3/2} M_c^{3/2} = 1$$

$$\text{Hayashi (243):} \quad E_c = 1.52 \tag{6.64}$$

$$\text{Graessley (227):} \quad 0.262 E_c^{5/2} = 1$$

$$\text{Graessley (212):} \quad 0.099 E_c^{5/2} = 1$$

in which E_c is M_c/M_e, the number of entanglements per molecule at the intersection.

The Bueche and Chikahisa relations reduce to roughly $M_c/M_e = 2$ and $M_c/M_e = 1$ respectively, when typical values of S^2/M and ϱ are inserted and

$K(s)$ is taken to be approximately unity. Hayashi's result is $M_c/M_e = 1.52$, while Graessley's expressions give values of 1.71 and 2.52 respectively. Thus, none of the friction factor theories is seriously inconsistent with the observed magnitude of M_c/M_e. The Bueche and Chikahisa formulas predict some residual variation of M_c/M_e with $(S^2/M)^{3/2}\varrho$ and the magnitude of M_c itself, while the Hayashi and Graessley results require M_c/M_e to be the same for all linear polymers. None of these results account satisfactorily for the detailed variations among polymers noted earlier.

A theoretical value for M_c'/M_c can be calculated by equating Graessley's expression for J_e^0 [Eq. (6–60)] with the Rouse expression:

$$0.4/vkT = 1.798/vE_ckT.\qquad(6.65)$$

The result is $E_c' = 1.798/0.4$, giving $M_c'/M_e = 4.5$. With $M_c/M_e = 2.52$ from the same theory, $M_c'/M_c = 1.8$ is obtained. The observed values are about twice this value; part of the difference is probably attributable to residual polydispersity in the experimental samples.

Thus, the available theories are consistent with the observation that $M_e < M_c < M_c'$. There is certainly no justification for the view sometime expressed that the characteristic molecular weights differ because different types of entanglements are responsible for the properties concerned.

It should be noted that the Bueche theory predicts a slightly different form for the variation of $(M_c)_{soln.}$ with concentration than that inferred earlier on the basis of experimental data:

$$c^4(M_c)_{soln.}^{7/2} = \varrho^4 M_c^{7/2}$$

or

$$(M_c)_{soln.} = \left(\frac{\varrho}{c}\right)^{8/7} M_c \quad \text{(Bueche theory)}.$$

Current experiments are not accurate enough to distinguish between this expression and Eq. (5.17). Finally, Bueche's theory is the only attempt to discuss η_0 vs M in the vicinity of M_c. Berry and Fox (16) have concluded that the predicted transition is too gradual. There is no theory for the behavior of J_e^0 in the vicinity of M_c'.

6.5.2. Relationship between the Characteristic Molecular Weights and Molecular Structure

Values of M_c (and M_e and M_c') are found to vary widely from one polymer to another, although the variations are much smaller (ca $\pm 50\%$) when comparisons are made in terms of n_c, the number of chain bonds per molecule at M_c. Fox and Allen (245) have pointed out that the variations are reduced still further (ca $\pm 30\%$) if values of S_c^2/v are compared, S_c^2 being the unperturbed mean square radius of gyration of molecules with molecular weight M_c, and v the volume of polymer per main chain bond. The parameter S_c^2/v comes directly from the Rouse form and the empirical correlating expression:

$$\eta_0 = \frac{\zeta_0}{6} X_c \left(\frac{X}{X_c} \right)^a \tag{6.66}$$

in which $a = 1$ for $X < X_c$ and $a = 3.4$ for $X > X_c$. This equation fits the experimental behavior of many linear polymers (16), X being the structure factor of the Rouse theory, nS^2v. For undiluted polymers, the parameter X reduces to S^2/v. Puzzling discrepancies remain however. For example, S_c^2/v for polyisoprene is more than twice as large as that for 1,4 polybutadiene, despite their similarity in backbone architecture.

Hoffmann has attempted to estimate the magnitude of M_e directly from the structure (246). He suggests that entanglements comprise a special class of intermolecular contacts in which one chain is looped tightly around another, and in which the four strands leading away from this contact are oriented such that the loop resists rapid deformations in the manner of a cross-link. His estimation of the concentration of such arrangements in polyethylene goes as follows.

A tight intermolecular loop around any methylene group requires the occupation of approximately 4 to 6 nearest neighbor positions by methylenes of the partner chain. The possibility of a similar occurrence with another chain at the same methylene group is negligible, and the possibilities on adjacent methylene groups are greatly reduced. As a result of such packing considerations, only about one methylene in three can be involved in a potential entanglement. The fraction of these which are effective depends on the vectors drawn from the junction to the next effective junction along each of the four strands.

According to Hoffman, these vectors must satisfy several conditions, each condition having a certain probability p of fulfillment.

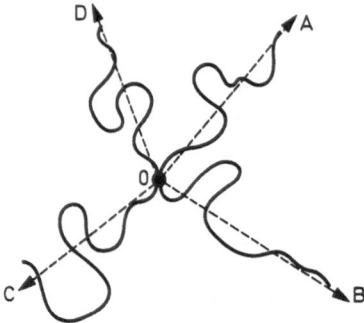

The angles AOB and COD should both be less than $90°$ ($p = \frac{1}{2} \cdot \frac{1}{2}$). The loop itself must lie on the correct side to engage when the chains are pulled in opposite directions ($p = \frac{1}{2}$). The bisector of angle COD must make a fairly large angle, say between $135°$ and $180°$, with the bisector of angle AOB $(p = \frac{1}{4}(2 - \sqrt{3}))$. The combined probability for all these events is therefore $(2 - \sqrt{3})/32 = 0.0084$, which is the fraction of potential loops which are effective. The number of methylene groups per effective couple is therefore approximately $3/0.0084 \approx 350$. Since there is an average of two strands per couple, the value of n_e is 175. Experimental values for n_e are in the range 100–250 for a variety of polymers.

The assumptions of the calculation are rather arbitrary, of course. The result perhaps only illustrates that the observed order of magnitude of n_e is not grossly inconsistent with the idea of topological restrictions by loops. The calculation gives no hint as to why the Fox-Allen parameter S^2/v should control the value of n_e or M_e however.

Other attempts to relate M_e to structure have not been successful. Tonelli (247) suggested that entanglement might be related to the probability of forming closed intramolecular loops along the chain. As an example, he estimated the probability p_{30}^{700} of forming at least one closed loop of greater than 30 main chain atoms in a sequence of 700 main chain atoms. No correlation between p_{30}^{700} and the experimental values of M_e or M_c was found in the six polymers used in the calculation. Indeed, attempts to relate entanglement effects strictly to the formation of transient intramolecular configurations or to local chain kinkiness or roughness (248) seem bound to fail, since they imply a much greater sensitivity to the details of local structure than is actually observed. Also, such transient structures would not contribute to equilibrium stress in crosslinked networks. The important contributions of entanglements to the equilibrium modulus, indeed the deduction of similar values of M_e in equilibrium and transient experiments (see Part 7), seems to rule out such kinetically controlled trapping mechanisms.

6.6. Effects of Polydispersity on η_0 and J_e^0

As noted earlier, η_0 in polydisperse systems appears to be a function of $\bar{M}_w$ alone. One problem in entanglement theories is how to introduce the strong molecular weight dependence of the entanglement friction factor without also introducing a much higher molecular weight average than $\bar{M}_w$ in polydisperse systems. Bueche (249) has applied his theory to a model polydisperse system, using a separate averaging process for ζ_e. He finds that viscosity depends on an average molecular weight $\bar{M}_t$ which lies between $\bar{M}_w$ and $\bar{M}_z$, being closer to $\bar{M}_w$ for small polydispersities and approaching $\bar{M}_z$ as polydispersity increases. Graessley (227) arrives at an average molecular weight of the form:

$$\bar{M}_t = \left[\int\limits_0^\infty M^{5/2} \, W(M) \, dM \cdot \int\limits_0^\infty M W(M) \, dM \right]^{2/7} \tag{6.67}$$

which also lies between $\bar{M}_w$ and $\bar{M}_z$. For a most probable distribution $\bar{M}_w : \bar{M}_t : \bar{M}_z$ as $1 : 1.23 : 1.50$. Both results proceed essentially from the combination of a separately averaged (communal) frictional coefficient with the ordinary Rouse analysis for polydisperse systems.

In a later examination of the problem, Graessley (*212*) considered the drag interactions between pairs of molecules in a uniformly effective medium, rather than interactions between individual molecules and a uniformly effective medium. The predicted viscosity in polydisperse systems was closer to $\bar{M}_w$ dependence than previously, although the sensitivity to high molecular weight components was still too large. The fault may lie with the assumption that all communal entanglement junctions, whether with large or small molecules, are equally effective as obstacles for the motion of the interaction pair. As a result of the assumption, pair interactions between the large molecules in a blend depend on their surroundings only through the total entanglement density of the system; the presumably reduced effectiveness as obstacles of their couples with small molecules is ignored.

Very little guidance from molecular theory is available on polydispersity effects in J_e^0. This of course is not surprising in view of the unsettled state of the theory of J_e^0 in monodisperse system. Recent theoretical calculations for entangled polymer blends (*212*) were found to be in good agreement with experiments when the blend components are not too different in molecular weight. However, as the difference increases, the theory systematically overestimates J_e^0 (*212, 250*). The same problem occurs as in the viscosity calculations: The contribution from pair interactions between large molecules is overestimated when the difference in size between the components is large.

7. Entanglements in Cross-linked Systems

Polymer systems can be converted to network-like structures by random crosslinking. Network formation renders permanent some of the large scale molecular arrangements which were present when the crosslinks were added. Topological arrangements, which can influence only rate processes in a system without crosslinks, become part of the permanent structure and may in principle affect its equilibrium mechanical properties as well. In particular, some portion of the entanglements which are present prior to crosslinking may be incorporated, being trapped by the permanent connectedness of the network strands and loops. If the system is subsequently deformed, the trapped entanglements may act as additional restraints and increase the equilibrium modulus beyond that conferred by chemical crosslinks alone.

Equilibrium moduli of networks can be used to distinguish to some extent among various physical conceptions of the entanglement interaction. For example, if entanglements were strictly transitory interactions with kinetically controlled lifetimes (*247, 248*), one would expect no contribution to the equilibrium properties. At the other extreme, the act of crosslinking could conceivably create as well as trap topological constraints (*251, 252*), resulting in an even higher entanglement contribution in the network than in the same system without crosslinks. Alternatively, the effects of crosslinking and pre-crosslinking entanglement might simply be additive (*253, 255*), with the equilibrium modulus controlled by the sum of elastically effective network strands between chemical crosslinks and a "trapped" fraction of the same entanglements that govern the plateau modulus.

Analysis of networks in terms of molecular structure relies heavily on the kinetic theory of rubber elasticity. Although the theory is very well established in broad outline, there remain some troublesome questions that plague its use in quantitative applications of the kind required here. The following section reviews these problems as they relate to the subject of entanglement.

7.1. Rubber-like Elasticity in Networks

The kinetic theory of rubber elasticity is so well known and exhaustively discussed (*17, 27, 256–257, 267*) that the remarks here will be confined to questions which relate only to its application in determining the concentration of elastically effective strands. In principle, both network swelling properties and elasticity measurements can provide information on network characteristics. However, swelling measurements require the evaluation of an additional parameter, the polymer-solvent interaction coefficient. They also involve examining the network in two states, one of which differs from its as-formed state. This raises some theoretical difficulties which will be discussed later. Questions on local non-uniformity in swelling (*17*) also complicate the inter-pretation. The results described here will therefore concern elasticity measurements alone.

In the original kinetic theory the network is composed of Gaussian chains or strands, joined at each end to junction points. Each strand is free to select any configuration which is consistent with the instantaneous positions of its junction points. The selection is made independently of the configurations of neighboring network strands. If the network is deformed, the mean positions of the junction points are assumed to move in an affine manner, *i.e.*, in proportion to the changes in macroscopic dimensions. The number of configurations available to each strand is altered by the resulting change in end-to-end coordinates. The result is a decrease in network entropy which gives rise to a mechanical restoring force.

If no volume or internal energy changes accompany deformation, the network is neo-Hookean (*258*), obeying the constitutive equation

$$p_{ij} = G \, C_{ij}^{-1} - P_0 \, \delta_{ij} \, . \tag{7.1}$$

The proportionality factor G is the equilibrium shear modulus of the network, given by

$$G = g v k T \, \frac{\langle r^2 \rangle}{\langle r_0^2 \rangle} \tag{7.2}$$

in which g is a parameter of order unity (which has continued to be the subject of some dispute), v is the number of network strands per unit volume, $\langle r^2 \rangle$ is the mean-square end-to-end separation of the network strands in the undeformed network, and $\langle r^2 \rangle_0$ is the mean-square end separation of the same strands with the junction-point constraints removed.

The "front factor" g as defined above[5] is unity in all the earlier theories (17). Recently Duiser and Staverman (233) have obtained $g = \frac{1}{2}$ and Imai and Gordon (259) $g = 0.54$ with Rouse model theories which make no *a priori* assumptions about the junction point locations after deformation. Edwards (260) also arrives at $g = \frac{1}{2}$ and Freed (261) deduces that $g = 1$ is an upper bound by similar approaches. The front factor usually assumed in the shifted relaxation theory of the plateau modulus is $g = 1$, although Chömpff and Duiser (232) obtain $g = \frac{1}{2}$ through their extension of the Duiser-Staverman result to entanglement networks. The physical reasons for the different values of g in different treatments are not clear at present.

James and Guth [see Ref. (17) for example] and more recently Eichinger (262) obtain $g = 1$, but claim that the progressive addition of crosslinks to form a network causes a reduction in chain dimensions through network shrinkage. In a sample calculation James and Guth obtain $\langle r^2 \rangle / \langle r^2 \rangle_0 = \frac{1}{2}$, although the implication of their suggestion is that the modulus should depend quite generally on the rate of crosslinking compared to the rate of configurational relaxation, and on the concentration of primary molecules in the crosslinking medium (263). Other authors assume that network strands are incorporated with equilibrium configurations and maintain the same configurations throughout the crosslinking process, $\langle r^2 \rangle / \langle r^2 \rangle_0$ being therefore unity for networks formed and tested under the same set of conditions.

A concomitant of coil shrinkage would seem to be a macroscopic shrinkage of the gel structure, called syneresis, if the crosslinking were carried out in solution. Syneresis is indeed observed at the later stages of crosslinking in highly diluted systems [see Refs. (17) and (257) for examples]. Some evidence of localized network inhomogeneities has also been adduced in polycondensation networks (264) and in solution-cured systems (265), which have been attributed to a localized syneresis. To this reviewer at least, appreciable coil shrinkage during random crosslinking of preformed primary chains in concentrated solutions or melts seems unlikely. It will be assumed that $\langle r^2 \rangle / \langle r^2 \rangle_0$ is unity in the lightly crosslinked, as-formed networks discussed here.

Most testing of the theory has been done with tensile deformations. According to Eqs. (7.1) and (7.2) the nominal tensile stress f, the tensile force per unit unstretched area, is related to α, the ratio of stretched to unstretched length, by:

$$f = g v k T \frac{\langle r^2 \rangle}{\langle r^2 \rangle_0} \left(\alpha - \frac{1}{\alpha^2} \right). \tag{7.3}$$

---

[5] Note that this definition of front factor differs from that given earlier by Tobolsky (162), since the term $\langle r^2 \rangle / \langle r^2 \rangle_0$ has been separated out.

Some aspects of the theory have been examined in detail. It is well established that the major part of the free energy change for small isothermal deformations is entropic. The small internal energy contribution ($\sim 10\%$ in the case of natural rubber, for example) is well accounted for by the temperature coefficient of $\langle r^2 \rangle$. Values of $d\ln\langle r^2\rangle/dT$ from elasticity measurements are in generally good agreement with values obtained in dilute solution (267), showing that intermolecular energy contributions must be rather small, at least in small deformations.

Finite extensibility of the network chains and volume changes induced by deformation also can influence the response. Quite understandably, the modulus increases as the chains become fully extended, but in networks with long network strands the upturn occurs only at rather large deformations ($\alpha > 2$). Small changes in volume most certainly do occur in tensile deformation since elastomers have roughly the same isothermal compressibility as small molecule liquids. However, Poisson's ratio is usually very close to 0.5, and, except for large deformations and in analyses to determine the internal energy contribution, the effect of volume changes is probably quite small (266).

Even when the above complications are negligible or properly accounted for and when strain-induced crystallization is absent, the stress-strain curves for networks seldom conform to Eq. (7.3). The ratio $f/(\alpha - 1/\alpha^2)$ generally decreases with elongation. An empirical extension of Eq. (7.1), the Mooney-Rivlin equation, has been used extensively to correlate experimental results:

$$p_{ij} = C_1 \, C_{ij}^{-1} - C_2 \, C_{ij} - P_0 \, \delta_{ij} \tag{7.4}$$

in which C_1 and C_2 are experimentally determined constants[6], typically of similar magnitude. For tensile deformations the stress based on initial area is

$$f = C_1\left(\alpha - \frac{1}{\alpha^2}\right) + C_2\left(1 - \frac{1}{\alpha^3}\right). \tag{7.5}$$

Even this form is probably only an approximation (267). More complicated expressions appear to be necessary to fit data on the same sample for different types of deformation (268). For sufficiently small deformations the tensile modulus becomes

$$\lim_{\alpha \to 1}\left\{\frac{f}{\alpha - 1}\right\} = 3(C_1 + C_2). \tag{7.6}$$

---

[6] Many authors use $2C_1$ and $2C_2$, reflecting a definition of the constants in terms of the elastic energy function. The factor of 2 is of course arbitrary and irrelevant to the discussion here.

Earlier the C_2 term was thought to reflect merely a lack of thermodynamic equilibrium in the measurements. Indeed, it is extremely difficult to achieve equilibrium in unswelled rubber networks (269). Although the magnitude of C_2 does appear to be sensitive to non-equilibrium effects and is reduced as equilibrium is approached, it now seems clear that many systems display finite C_2 values even at equilibrium (17). The temperature dependences of C_1 and C_2 are quite similar (267, 270), suggesting that both have essentially entropic origins. On the other hand, diluents influence C_1 and C_2 rather differently. The C_2 term generally decreases rapidly with swelling. When compared at constant swelling ratio, the values of C_2 obtained are remarkably insensitive to the thermodynamic nature of the solvent (271). The magnitude of C_2 does depend on the type of curing system used however (267). On the other hand, C_1 changes in accord with the predictions of kinetic theory. Thus, its value depends mainly on the crosslink density and is altered by swelling in the manner required by kinetic theory:

$$C_1 \sim v \, \frac{\langle r^2 \rangle}{\langle r^2 \rangle_0} \propto \left(\frac{V_0}{V} \right) \left(\frac{V}{V_0} \right)^{2/3} = \varphi^{1/3} \tag{7.7}$$

in which V_0 is the volume of the sample when the crosslinks were introduced, V is the swollen volume, and φ is the volume fraction of polymer in the swollen state V_0/V. In a recent study of peroxide-cured natural rubber (270) the quantity $C_1 \varphi^{-1/3}$ was independent of swelling ratio, while $C_2 \varphi^{-1/3}$, which was approximately $0.4 C_1$ in the dry state, decreased linearly with φ and extrapolated to zero for $\varphi \approx 0.2$. Sulfur cured vulcanizates of natural rubber and various styrene-butadiene rubbers behaved in an almost identical manner (271). A continuum explanation has been advanced for the rapid reduction of C_2 with swelling (272), but Flory and Mark (273) have pointed out the ambiguities inherent in such an approach.

For the same polymer and curing procedure, networks formed in the presence of various diluents ($\varphi \approx 0.15 - 0.40$) give practically no C_2 contribution when tests are made in the dry state (274). Kinetic theory requires $C_1 \varphi^{-2/3}$ to be constant in this case (φ being the volume fraction of polymer during solution cure), assuming comparable numbers of crosslinks per primary molecule are introduced. Experimentally $C_1 \varphi^{-2/3}$ is close to the values of C_1 obtained for dry state cures, although they do decrease somewhat in the more dilute curing systems ($\varphi < 0.2$). Solution cured polydimethyl siloxane networks show similar behavior in their dry state moduli (275).

Gent and Rivlin (276) have suggested that C_1 is in fact the kinetic theory term, related directly to the concentration of effective network strands bounded by chemical crosslinks. The molecular origin of the C_2 term is unknown, although many possibilities have been suggested (17, 267, 277). The above experiments seem to rule out both isolated chain explanations, such as intramolecular excluded volume, and suppositions based on local ordering of chains. If either were important, it should logically contribute in dry state measurements, regardless of whether the network had been formed originally in the dry state or in solution. Non-Gaussian chain behavior seems an unlikely explanation also, since C_2 apparently is small in *both* the extended (swelled) and supercoiled (dry-state after solution cure) states of the network strands.

Various authors have suggested that the source of deviations from neo-Hookean behavior lies in the large scale organization of the network itself (*257, 267, 274*). An origin in network topology would explain both the sensitivity of C_2 to the conditions during network formation and the observations that C_2, like C_1, is largely of entropic origin. Alfrey and Lloyd (*252*) have emphasized that the conventional kinetic theory deals with an assembly of strands whose configurations are statistically independent, in effect with "ghost" chains which can pass freely through the backbone contours of neighboring chains. Chains in real networks must certainly be subject to topological restrictions, and the configurations of neighboring chains must therefore be conditionally dependent. If such dependency affects the number of configurations differently in the deformed and undeformed states, then the entropy of deformation must contain an extra contribution. This extra contribution, which one might call the topological contribution to distinguish it from the independent strand term of the conventional theory, could produce a response different than that of a neo-Hookean solid. In particular, it might contribute the C_2 term and some portion of C_1.

If this view is correct, the measurements of the equilibrium modulus can provide some information on network topology, provided the independent strand contribution can be evaluated separately. *Principal attention should focus on the modulus for small deformations.* Since the structure is essentially undisturbed by such measurements, the small deformation modulus should contain the most direct molecular information. Proceeding beyond this to an interpretation of large deformation behavior would be expected to present a comparable increase in difficulty to that encountered in moving from molecular interpretation of η_0 to molecular interpretation of η vs. $\dot{\gamma}$ behavior in melts and concentrated solutions. As a corollary, *the mechanical behavior of the network should be evaluated in the same state of swell as that in which it was formed.* Swelling or supercoiling subsequent to network formation itself involves a finite deformation of the network and under these circumstances even small deformation moduli may not be easily interpretable in terms of topological contributions. Indeed, assuming that the term represented by C_2 contains an important part of the topological information, the unexplained changes in C_2 with swelling or supercoiling suggest a relative change in the independent strand and topological contributions. Of course, a complete theory of topological effects must ultimately be able to account for swelling and large deformation behavior, but until then, interpretation seems necessarily limited to small deformation behavior in networks at the same state of swell as when the crosslinks were added. If this is the undiluted state, as is many times the case, then there is the corresponding, extremely difficult, experimental problem of achieving stress equilibrium in the modulus measurement.

7.2. Tests of Theoretical Modulus Values—Model Networks

Quantitative tests of the absolute modulus value from kinetic theory [Eq. (7.2)] are quite scarce; independent determination of the network strand concentration is generally the difficulty. Schaefgen and Flory (*278*) prepared model polyamide networks by coupling pre-formed star branched molecules. The modulus was accurately proportional to the concentration of network strands, but the values exceeded the theoretical moduli (calculated as $G = vkT$) by approximately a factor of 2, implying $g = 2$ for these networks. The molecular weights of the strands (8000–20000) probably exceed M_e for polyamides to some extent, so g may be too high because of additive entanglement contributions.

In two other studies (279, 280) model networks were formed by the end-coupling of pre-formed linear chains with multifunctional coupling agents. Strand molecular weights in both cases were of the order of M_e for the non-crosslinked system; topological contributions to the modulus might be expected to be smaller in these networks than in systems with longer strands, although not necessarily absent. Kraus and Moczygemba (279) coupled carboxy-terminated polybutadiene chains ($M = 5500$) to form a network in the dry state. With an excess of coupling agent they observed modulus values approaching $C_1 = 4.8 \times 10^6$ dyne/cm^2 and $C_2 \approx 0$ at 25 °C. The value of vkT for this composition is 4.1×10^6, implying $g = 1.2$. The authors suggested that entanglement effects should be negligible in these networks because the molecular weight of the strands is smaller than M_c ($M_c = 5900$ for 1,4 polybutadiene). However, M_e is only 1900 for this polymer, so it is not clear that entanglement contributions are absent.

Prins and co-workers (280) reported moduli on polystyrene networks which had been formed by adding small amounts of divinyl benzene to preformed polystyryl dianions. The gels were dried, reswelled to equilibrium in benzene and tested in simple compression. The equation for G in this case becomes:

$$G = g v^* k T \, \varphi_0^{2/3} \, \varphi^{1/3} \tag{7.8}$$

in which v^* is the network chain concentration in the dry state, φ_0 is the volume fraction of polymer during network formation, and φ is the volume fraction during modulus measurement. These molecular weights are indeed much smaller than $(M_e)_{\text{soln}}$ for polystyrene at the curing and swelled-state concentrations. Gels were prepared with network strand molecular weights from 7200 to 18700. Deduced values of g ranged from 0.4 to 0.55; no information on C_2 was reported.

Allen and co-workers (280a) have recently prepared and studied a series of model polystyrene networks. These networks were formed from narrow distribution polystyrenes ($M = 70000$–240000) containing small but known amounts of secondary amine side groups. The amines were coupled in solution by reaction with diisocyanates, resulting in networks with known values of v_c ($M^* = 7000$–17000). Moduli were determined for the networks at the same diluent concentrations as used to form the networks ($\varphi = 0.05$–0.25). After applying approximate corrections for intramolecular (sterile) crosslinks and trapped entanglement contributions, the authors concluded that a value $g = 0.5$ would satisfy most of the data. Recent theoretical calculations by Tonelli and Helfand (280b) suggest that sterile loops may play some role even for networks formed in the undiluted state. On the other hand, when gel-sol studies are used to evaluate crosslinking densities (see below), sterile loops remain uncounted, so there is at least partial compensation for their reduced effectiveness in equilibrium mechanical properties.

7.3. Tests of Theoretical Modulus Values—Statistical Networks

Networks formed by random multifunctional condensations or random cross-linking can be characterized by a combination of gelation studies and statistical calculations of structure, based on the equal reactivity principle (281). There is still some debate, however, about the correct counting procedure for calculating the concentration of elastically effective network strands. Flory (3) noted that, in a gel composed of N primary chains and C tetrafunctional crosslinks, $N-1$ of the crosslinks are required merely to join the primary chains together. Each crosslink beyond this number links two parts of the structure together, creating in the process two additional closed loops. According to Flory, each such closed loop corresponds to an elastically effective strand. Thus, the total number of active strands is

$$Flory\ criterion: \quad N_A = 2(C - N + 1) \approx 2(C - N). \qquad (7.9)$$

Alternatively, N_A is the number of mers participating in crosslinks ($2C$) minus the number of chain ends ($2N$). Tobolsky (162) has criticized the Flory criterion, showing that Eq. (7.9) is incorrect near the gel point. The Flory criterion has been used in most of the network analyses, however, and for purposes of comparison we will carry it along.

Scanlan has suggested another criterion (282). An effective network junction point is a crosslink in which at least three of the four strands radiating from it lead independently to the network. A crosslink with only two strands anchored to the network simply continues an active strand; a crosslink with only one anchored strand is part of a dangling end and can make no elastic contribution at equilibrium. An elastically effective strand is therefore one which joins two effective network junction points. Accordingly, the total number of active strands is simply one half the number of gel-anchored strands radiating from effective junction points:

$$Scanlan\ criterion: \quad N_A = \frac{C}{2}(3p_3 + 4p_4) \qquad (7.10)$$

in which p_3 and p_4 are the fractions of crosslinks with 3 and 4 gel-anchored strands.

The two criteria give somewhat different results for the effective strand concentration. Consider, for example, the random addition of C crosslinks to

ν_0 monodisperse primary chains per unit volume (primary chain length $= P$ repeating units). The gel fraction w_g is related to crosslinking index γ through the equation (valid for $P \gg 1$) (27, 33):

$$1 - w_g = e^{-\gamma w_g} \tag{7.11}$$

in which γ is $2 C/\nu_0$, ν_0 is ϱ/M, ϱ is the polymer density and M is the primary chain molecular weight $m_0 P$. The crosslinking index of the gel is higher than the overall crosslinking index (27):

$$\gamma_g = (2 - w_g) \gamma . \tag{7.12}$$

A simple calculation of the concentration of active strands, following Langley (255), yields

$$\text{Flory criterion:} \quad \nu_c = \nu_0 \left[(2 - w_g) \ln \frac{1}{1 - w_g} - 2 w_g \right] . \tag{7.13}$$

Scanlan's criterion has been used extensively by Gordon and co-workers. The resulting equation for monodisperse chains, again for $P \gg 1$, is (283):

Scanlan criterion:

$$\nu_c = \nu_0 \left[(2 - w_g) \ln \frac{1}{1 - w_g} - 2 w_g \right] \left[2 + \frac{\frac{1}{1 - w_g}}{\ln \frac{1}{1 - w_g}} - \frac{1}{w_g} \right] . \tag{7.14}$$

Table 7.1 compares ν_c/ν_0 according to the two criteria. Values for primary chains with exponential distribution ($M_w/M_n = 2$) are also included. The number of active strands according to the Scanlan criterion is larger by a factor of 1.5 at the gel point. The ratio decreases slowly with increasing γ and finally approaches a value of unity for large numbers of crosslinks per primary chain. Both expressions approach the form $\nu_c = \nu_0 \gamma$ for sufficiently large values of γ.

The Flory criterion requires that each crosslink between units which are already part of the gel must increase the number of active strands by 2. Examination of example structures shows that this number according to Scanlan's criterion is either 1, 2, or 3, depending on the structures involved. The difference between criteria is therefore a real one in physical terms. In the author's opinion, the Scanlan criterion is clearly the correct one. Any other choice would seem to depend ultimately on such questions as whether junctions with three anchors in the gel are equivalent to 4-anchor junctions insofar as the assumption of affine displacement is concerned. To the extent that the strands behave as linear

Table 7.1. Concentrations of elastically effective strands according to the Flory and Scanlan
criteria for random crosslinking of monodisperse primary chains

γ	w_g	v_c/v_0 (Flory)	v_c (Scanlan)/v_c (Flory)	
1	0 (gel point)	0	3/2	$(3/2)^a$
1.5	0.59	0.08	1.42	(1.40)
2.0	0.80	0.33	1.37	(1.31)
3.0	0.94	1.2	1.29	(1.19)
7.0	0.999	5.0	1.15	(1.03)
∞	1	∞	1	1

[a] Values in parentheses were obtained at the same gel fraction w_g by calculations for
primary chains with an exponential distribution of lengths.

springs (conform to the Gaussian distribution) junctions of any functionality
greater than two would seem to be equivalent.

Tobolsky and co-workers (283a) have determined g for a large number of
networks prepared by copolymerizing various acrylate and methacrylate mono-
mers with small amounts of a tetrafunctional methacrylate monomer. Essentially
complete conversion was achieved in all cases; the amount of extractibles was
found to be negligible. Two effective strands were assumed to result from each
tetrafunctional mer in the system. The calculated molecular weights between
junction points ranged from 12000 to 50000 ($v_c \approx 2 - 8 \times 10^{-5}$ moles/cc). This is
probably below M_e for the chains with large pendant ester groups, but comparable
to or above that for the methyl and ethyl esters. Values of g from tensile moduli
varied systematically from 1.5–2.0 for the methyl and ethyl ester chains (consistent
with some entanglement contribution) to 0.2–0.3 for chains with bulky pendant
groups. The authors speculated that crowding at the junction points may have
influenced network structure in the latter case, resulting in fewer than the
theoretical number of effective strands and hence in g values which are too low.
It is difficult to draw clear conclusions from these results, aside from noting that
$g \approx 0.5$–0.7 appears to satisfy most of the data from chains in which the pendant
groups are not too bulky, but for which the strand lengths are shorter than the
entanglement spacings.

Gordon et al. (284) have measured moduli on networks formed by con-
densing decamethylene glycol and benzene- 1, 3, 5 triacetic acid. The concentration of active strands in each sample was calculated from gel point data
and the statistical theory of structure appropriate to this system, using Scanlan's
criterion. They found rather good agreement between the initial modulus and
that calculated using Eq.(7.2) with $g = 1$ for networks just beyond the gel point.
In this region the gel is quite open and theoretically (see following section) the
number of trapped entanglement is small relative to the number of strands
bounded by chemical crosslinks. At higher extents of reaction the moduli became
systematically higher than values predicted similarly.

To the author's knowledge no other evaluations of the g factor have been
made on statistical networks under circumstances in which the contribution of
trapped entanglements would be expected to be negligible. Aside from Gordon's

result of $g \approx 1$, the data on both model (280, 280a) and statistical (283a) networks suggest very strongly that $g < 1$, and in fact that g probably lies in the vicinity of 0.5 for many systems. Further estimates of g have been made in situations where entanglements are important but separable, at least theoretically, in their effects on the initial modulus (see following Section 7.4).

7.4. Determination of Entanglement Density in Networks

7.4.1. Theories of Entanglement Trapping

Mullins (253) and Bueche (254) suggest that entanglements which are trapped in the system during crosslinking will simply add to the number of active strands in the rubber elasticity modulus. An entanglement site on a primary chain is considered to be trapped only if it occurs between two crosslinking sites on that chain. For a network formed by adding C crosslinks to N primary chains of (number-average) degree of polymerization P, the number of crosslinked units in the system is $2C$. The probability that a repeating unit selected at random is contained on a strand bounded at both ends by crosslinks $(C \gg N)$ is $\left(\dfrac{2C}{2C+N}\right)^2$. With C large compared to N, the gel fraction is practically unity, and for the network, $\left(\dfrac{2C}{2C+N}\right)^2 \approx 1 - \dfrac{N}{C}$. From a total of EN entanglement sites, the number which are trapped is therefore $EN(1 - N/C)$. With Flory's criterion, the total number of active network strands according to Bueche and Mullins is

$$N_T = 2C - 2N + EN\left(1 - \frac{N}{C}\right). \tag{7.15}$$

When cast in terms of strand concentrations, the Bueche-Mullins equation becomes

$$v = (v_c + v_e)\left(1 - \frac{2v_0}{v_c}\right) \tag{7.16}$$

in which v_c is $v_0 \gamma$, the concentration of repeating units which participate in crosslinks. The primary chain concentration v_0 is ϱ/M, in which M is the (number-average) molecular weight of the primary chains. Finally v_e is ϱ/M_e, M_e being the molecular weight between entanglement points, a property which is assumed to be independent of both primary chain length and crosslink density. The Bueche-Mullins trapping factor is therefore $(1 - 2v_0/v_c)$.

Equation (7.16) rearranges to give

$$v = v_c + v_e - (1 + v_e/v_c)\frac{2\varrho}{M}. \tag{7.17}$$

The active strand concentration v is obtained from measurements of the initial modulus using Eq.(7.2) with $g = 1$. Values of v for samples at the same crosslink density but with different primary molecular weights are extrapolated to $1/M = 0$, giving $v_c + v_e$. Values of $v_c + v_e$ obtained at two or more crosslink densities provide v_e, and hence M_e, by extrapolation. An advantage of this method is that only relative values of crosslink density must be known: absolute values of v_c are not required. If absolute values of v_c are known, the g factor can be evaluated as well.

An intuitively more satisfying requirement for trapping was suggested by Ferry and co-workers (285). In order for any entanglement to contribute to the equilibrium modulus, all four strands leading from that entanglement must terminate in crosslinks. To the same approximation as the Bueche-Mullins relation ($C \gg N$), and with the Flory active strand criterion, the strand concentration becomes

$$v = v_c(1 - 2v_0/v_c) + v_e(1 - 2v_0/v_c)^2 . \tag{7.18}$$

The Ferry trapping factor is therefore $(1 - 2v_0/v_c)^2$. Unfortunately, the simple extrapolation procedure of Kraus (286) can no longer be applied, but numerical search techniques should still suffice to determine M_e.

Bueche-Mullins trap   Ferry trap   Langley trap

Both the Bueche-Mullins and the Ferry relations are attempts to allow for the effect of finite primary chain length on active strand and entanglement strand concentrations in well developed networks. In applying such relations to determine M_e in networks, it is clearly desirable both to begin with very long primary chains and to measure moduli on networks which have several cross-linked units per primary chain. Thus, if $M \gg M^* \gg M_e$, where M^* is the average molecular weight between crosslink junctions, the trapping factor is essentially unity, and $v \approx v_c + v_e \approx v_e$. The uncertainties due to the chain end correction to the use of the Flory criterion for active strands, and to independently estimated values of v_c (or extrapolation to $v_c = 0$). will be small under these conditions.

Langley (255) has developed a refined definition of the Ferry trapping factor, and he also accounts theoretically for effects of the primary chain distribution and concomitant random chain scission during the crosslinking process. According to Langley, an entangling interaction between chains contributes to the equilibrium modulus only if all four strands leading away from the entanglement junction have their own separate connections to the network. It is assumed that all repeating units in the system have the same probability of participating in an entanglement. The fraction of trapped entanglement couples T_e is then p^2, where p is the probability that each of the two strands leading away from a randomly selected repeating unit in the system is anchored separately to the network. For example, in a randomly crosslinked system of monodisperse primary chains, the trapping factor is (255):

$$T_e = \left[2 - w_g - \frac{2w_g}{\ln \frac{1}{1-w_g}} \right]^2. \tag{7.19}$$

Langley employs the Flory criterion to evaluate the number of active strands bounded by chemical crosslinks. The total concentration of active strands, including strands bounded by trapped entanglement junctions, can be expressed (255)

$$v = v_0 \gamma w_g T_e^{1/2} + v_e T_e. \tag{7.20}$$

Equation (7.20) is general, although the expression for T_e itself depends on the primary distribution, the gel fraction, and the relative rates of random crosslinking and chain scission.

With the Gordon equations (273) one can easily extend Langley's analysis to the Scanlan criterion for active strands bounded by chemical crosslinks. With the same generality as that of Eq. (7.20):

$$v = \frac{v_0 \gamma}{2} T_e^{1/2}(3w_g - T_e^{1/2}) + v_e T_e. \tag{7.21}$$

For either Eq. (7.20) or (7.21) the trapping factor T_e can be calculated from a rearranged and simplified version of Eq. (17) in Ref. (255):

$$T_e = \left[2x - w_g - \frac{2x}{yP_n}(1 - F_0) \right]^2, \tag{7.22}$$

in which

$$F_0 = \sum_{P=1}^{\infty} N(P) e^{-yP}. \tag{7.23}$$

The initial number-average degree of polymerization of the primary chains is $\bar{P}_n$, $N(P)$ is the fraction of primary chains with P repeating units, and x and y have the meanings given by Langley (255):

$$x = qw_g/(qw_g + p), \tag{7.24}$$

$$y = qw_g + p \tag{7.25}$$

where q and p are the respective fractions of repeating units which participate directly in crosslinks and which have undergone scission.

Application of Eq.(7.20) or (7.21) to modulus measurements provides both a value of M_e and a value of the front factor g. For example, the combination of Eqs.(7.20) and (7.2) leads to:

$$\left(\frac{\langle r^2 \rangle_0}{\langle r^2 \rangle}\right)\left(\frac{G}{RT}\right)\frac{1}{T_e} = \frac{gv_0\gamma w_g}{T_e^{1/2}} + \frac{g\varrho}{M_e} \tag{7.26}$$

in which v_0 is expressed in moles/volume. The modulus G for a series of samples crosslinked to different extents is obtained $(G = f/(\alpha - 1/\alpha^2)$ at small deformations), and the quantities γ, w_g, and T_e are evaluated for each network from the gelation behavior of the system and its primary chain distribution. The quantity $\langle r^2 \rangle/\langle r^2 \rangle_0$ is unity if the modulus is measured at the same state of swell as is present during crosslinking. Thus g and M_e are separately determinable from the slope and intercept of a plot of G/T_e vs $\gamma w_g/T_e^{1/2}$. Application of Eq. (7.21) proceeds similarly.

It should perhaps be emphasized that the trapping relations themselves can be stated in more general terms than those implied by interpenetrating loop ideas. The Langley relation is simply the fraction of intermolecular contacts between mers in which both are parts of active strands after crosslinking. It is assumed to measure the fraction of pair-wise topological arrangements in the system which are rendered permanent when crosslinks are added, and therefore the fraction of the total potential topological contribution which joins with the chemical strand contribution to determine the equilibrium modulus. Viewed thusly, the trapping analysis depends not at all on specific models of the entanglement interaction. The values of M_e deduced from small deformation network elasticity have the same meaning in principle as those from the plateau modulus: when described in terms of localized interactions, the topological contribution is equivalent to that of a permanent network with crosslink spacing M_e. Comparisons between values of M_e from network elasticity and from the plateau modulus of uncrosslinked polymers provides therefore a clear and model-free test of the topological entanglement concept.

7.4.2. Experimental Results

The Bueche-Mullins method has been applied in the separation of the modulus contributions of crosslinks and entanglements in several elastomers. A front factor of $g = 1$ was then used to determine M_e. The Langley method has also been applied in a few cases, resulting in values of both g and M_e. These works are summarized below; results are collected in Table 7.2.

Kraus (286) obtained $M_e = 6000$ for peroxide-cured 1,4 polybutadiene networks formed from samples of different primary molecular weights. Moduli were determined indirectly from swelling behavior, previously correlated, however, with elastic properties. Chain scission was assumed to be absent and the chemical crosslinking densities were obtained indirectly. Van der Hoff and Buchler (277) found $M_e = 4300$ for 1,4 polybutadiene networks which were cured to different extents by a quantitative crosslinking agent at room temperature. Both studies used the Bueche-Mullins method. Pearson et al. (287) have applied Langley's method [Eq. (7.26)] in the analysis of radiation-crosslinked 1,4 polybutadiene, obtaining $g = 1.22$ and $M_e = 5900$ from gelation data and moduli of networks which were crosslinked dry and subsequently swollen in toluene.

Vander Hoff and Buchler (277) found $M_e = 13000$ for natural rubber by applying the Bueche-Mullins analysis to the data of Bristow (288). Examination of the natural rubber results published some years ago by Flory et al. (289) yields $M_e \approx 15000$. The data of Mullins (290) provides a value of 10000 for 1,4 polyisoprene if the total initial modulus is used in the analysis (Mullins reported $M_e = 16000$ by examining the effects of primary molecular weight and extent of cure on C_1 alone; the initial modulus G is $C_1 + C_2$ and $C_2 \approx 0.6 C_1$ in the samples examined).

Bueche and co-workers (253, 291) obtained $M_e = 13000$ in radiation-cured poly(dimethyl siloxane) networks. They also determined absolute values of the chemical crosslink density, using molecular weight measurements in the pre-gel region to establish a crosslink density vs dose calibration. The front factor for the chemical crosslink contribution was found to be $g = 0.65$ (291). Langley and Polmanteer (292) have recently applied Langley's method to three well characterized samples of poly(dimethyl siloxane). Gel fractions and moduli were measured in radiation-cured networks. The front factor g was found to be 0.81, and the molecular weight between entanglement points (using $g = 0.81$) was $M_e = 8600$ (Fig. 7.1). These workers have also analyzed the same experimental data using the Scanlan active strand criterion [Eq. (7.21)], finding in this case $M_e = 5600$ and $g = 0.60$ (293). In an earlier study of poly(dimethyl siloxane) Langley and Ferry (294) found that the fraction of entanglement strands which are trapped in the structure is given with good accuracy by the calculated trapping factor T_e. The number of active strands was determined by both modulus and swelling measurements (the latter calibrated, however, in terms of modulus values), and then calculated theoretically from gelation data and the primary chain distribution, using the Langley equations.

Ferry and co-workers (295) have obtained a value $M_e = 13000$ for 1,2 poly-butadiene by a new method which involves crosslinking the polymer in a strained state. The initial extension of the uncrosslinked sample, the recovery after

Table 7.2. Values for the molecular weight between entanglements M_e from the equilibrium properties of crosslinked networks

Polymer	M_e from the viscoelastic plateau	M_e from equilibrium network modulus	Value of g factor obtained	Method	Investigators
1,4 Polybutadiene	1900	6000		Bueche-Mullins	Kraus (286)
		4300		Bueche-Mullins	Vander Hoff and Buchler (277)
		5900	1.22	Langley (Flory criterion)	Pearson et al. (287)
1,4 Polyisoprene and natural rubber	5800	13000		Bueche-Mullins	Vander Hoff and Buchler (277)
		15000 (est.)		Bueche-Mullins	Data of Flory et al. (289)
		10000		Bueche-Mullins	Data of Mullins (290)
Polydimethyl siloxane	8100	13000	0.65	Bueche-Mullins	Bueche and co-workers (254, 291)
		8600	0.81	Langley (Flory criterion)	Langley and Polmanteer (292)
		5600	0.60	Langley (Scanlan criterion)	Langley and Polmanteer (292)
1,2 Polybutadiene	3550	4100		Kramer-Ferry	Ferry and co-workers (295, 296)

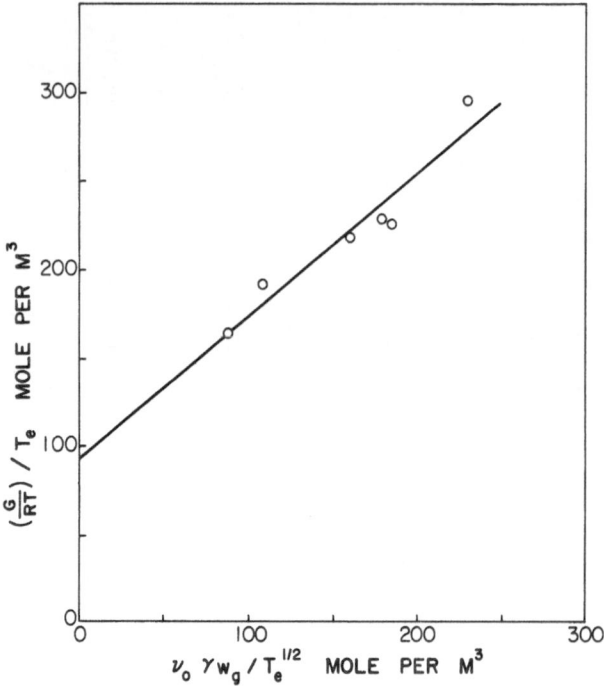

Fig. 7.1. Plot of modulus data on poly(dimethyl siloxane) networks to determine g and M_e by Langley's method [Eq.(7.26)] (292)

crosslinking, and the modulus of the polymer in its final rest state were measured, and the entanglement density in the original state was calculated using Flory's theory of composite Gaussian networks (223). The above value appears to have been somewhat high because of entanglement slippage during the initial deformation. Recent results, obtained with lower temperatures during deformation and crosslinking, yield $M_e = 4100$ (296), in good agreement with the plateau modulus value [$M_e = 3550$ (15)].

Baldwin and Ver Strate (297) employed expressions equivalent to Eqs.(7.20) and (7.21) to analyze modulus data on ethylene-propylene ethylidene norbornene terpolymers swelled with a solvent. They fixed the crosslinking density by postulating complete utilization of the peroxide curative added, this justified on the basis of considerable ancillary work. The deduced molecular weight between entanglements (assuming $g = 1$) on networks swollen to equilibrium was found to depend on the chemical crosslink density, with values ranging from $M_e = 2300$ at low crosslink densities to 1400 at higher values. The order of magnitude of M_e is probably correct. However, the systematic increase in M_e with crosslinking density may have been caused by the fact that the moduli were measured on networks which swell to different extents at equlibrium. As pointed out earlier, entanglement contributions to the modulus might be expected to decrease with increasing swelling ratio, which would produce just the trend observed.

Table 7.2 shows that the reported values of M_e from equilibrium moduli of networks are consistently larger than those obtained from the plateau modulus using $g = 1$. Superficially the results imply that the entanglement density affecting equilibrium properties is somewhat less than that affecting dynamic properties such as the plateau modulus. However, the choice of active strand and trapping criteria is clearly important in the numerical value of M_e and g obtained. In particular, M_e deduced by the Bueche-Mullins method would be expected to be systematically larger than that from the more accurate Langley method because the former overestimates the number of trapped entanglements. It may be significant that the results of Langley and Polmanteer (292) and Langley and Ferry (294) with poly(dimethyl siloxane) are the most precise in terms of characterization of primary chains and gelation behavior, and that both provide equilibrium values of M_e which agree best with values from the plateau modulus. The analysis of 1,4 polybutadiene by Pearson et al. (287), is comparable in precision and yields M_e somewhat larger than that from the plateau modulus. However, the mechanical properties in this latter study were determined in the swollen state. By inference from the sensitivity of C_2 to swelling noted earlier, the effects of network topology on elasticity are reduced in swollen systems. The Pearson study might therefore be expected to produce values of M_e which are too large.

It is clear that the application of Langley's method in other polymer systems is essential to settle questions about M_e and g in networks satisfactorily. The Ferry composite network method (223, 296) appears to be broadly applicable as well, although requiring special care to minimize slippage prior to introduction of the permanent crosslinks. (One is also still faced with the difficult question of whether g is the same for entanglements in crosslinked networks and in the plateau region of dynamic response.) Based on the limited results of these two methods in unswelled systems, M_e values deduced by equilibrium and dynamic response appear to be practically the same.

7.5. Relaxation Processes in Networks

Stress relaxation in networks formed by random crosslinking tends to be spread over many decades in the time scale. Ferry (15) attributes the slow relaxation processes to dangling structures, composed perhaps of several primary molecules which are linked to the main body of the network in one or only a few places. The statistical theory of crosslinking predicts the existence of such structures. They should be most prevalent in the vicinity of the gel point, but some will be present even when there is an average of many cross-links per molecule. Relaxation is much more rapid in model networks which have no dangling structures and in networks swollen by solvent. Quantitative interpretation of these slow relaxations awaits a clearer picture of relaxation in branched skeletal structures. Recent work by Cohen and Tschoegl (298) on block copolymers with dangling ends suggests that such materials may provide models for the systematic study of slow relaxations in networks.

Ferry and co-workers (*299*) have recently reported linear viscoelastic studies on polyisobutylene networks containing linear polyisobutylene molecules which interpenetrate the network but are not chemically bound to it. Entanglement and crosslink spacings are roughly the same. In this case the unbound molecules relax in the network environment, in a manner presumed to be roughly comparable to the reptation model of De Gennes (*225*). One surprising aspect of this work is that the relaxation spectrum of the linear chain-network composite is roughly independent of the concentration of unbound chains, up to a weight fraction of 0.50. One might hope that in the future such systems will provide insight on the large-scale motions of individual chains in entangled or permanently connected structures.

7.6. Theoretical Analysis of Topological Constraints

7.6.1. Effects of Topological Classification

Edwards (*300*) considers the effect of entanglement in permanent networks to arise from the topological classification imposed on the strands when the network is formed. In the conventional theory of rubber elasticity, the network strands are free to choose configurations without regard to the configurations of neighboring strands. In real networks the strands cannot pass through one another's backbone contours. Thus, the configurations available to any strand depend on its topological relationship with neighboring strands; certain configurations become unavailable to each class which would be available to independent strands. These additional constraints have no effect on the configurational entropy of the network in its as-formed state, since the cross-links simply bind the chains permanently into what is already an equilibrium distribution of configurations. If the network is deformed or swelled, however, the additional constraints become important, as shown below, and the entropy change will differ from that for a network of independent strands.

The simplest example of topological classification is the knotting of individual strands, *i.e.*, self-entanglement. However, the effects attributed to entanglement in non-crosslinked polymers are clearly intermolecular. The simplest such case is that of pair-wise classification without self-entanglement. Consider the following three examples of strand pairs:

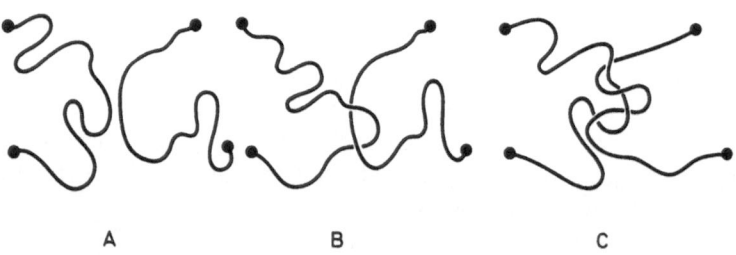

A   B   C

The equilibrium positions of the four junction points define the spatial relationship between the strands of each pair. For simplicity, the mean relative positions (internal coordinates) of the junction points of each pair are taken to be the same. The junction points of each strand are separately anchored to the network by at least two of their remaining strands, so each is an elastically effective strand according to Scanlan's criterion. The network itself in effect completes the loop for each strand, making the A, B, and C pairs as structurally distinct as catenane molecules (301).

Each pair is one example of a myriad of permanent topological classes for strand pairs whose spatial positions, described by the internal junction coordinates only, are the same, but whose structures cannot be transformed from one to the other without chain breakage. Members of each class have configurations available which are not available to members of other classes. Moreover, every distinguishable configuration of a strand pair must belong to only one such class. Considering pair-wise relationships only, the same topological classes will be available for any pair of strands in the network. However, the probability (relative population) of each class will depend strongly on the internal coordinates of their junction points, approaching a probability of unity for the unlooped class in widely separated pairs. One would expect that the apportionment among classes for pairs with the same internal coordinates would depend on the latent, *i.e.*, equilibrium, topological arrangement of the primary chains at the moment of crosslinking.

If the network is deformed, the mean internal coordinates of each strand pair assume new values, and the number of configurations available to pairs of each class changes. The number of configurations available to each class with the new internal coordinates will be proportional to the population of that class for pairs which form at equilibrium with this set of internal coordinates. The change in entropy with deformation is calculated from the changes in the number of distinguishable configurations of the strands, taking into account the permanence of topological classification, through the Boltzmann equation. The contribution of entanglements to the modulus is then obtainable from the difference between this value and the entropy of deformation for independent strands.

Edwards (300) was able to make some progress on the problem by showing that topological classification can serve only to raise the modulus. As a simple example, consider all strand pairs in a network which have, within some small tolerance, a specified set of internal junction coordinates. Suppose there are B such pairs, and that the strands of each pair, labeled 1 and 2, have ω_1 and ω_2 distinguishable configurations each as free strands, and fractions $(g_1)_0$ and $(g_2)_0$ respectively which have the end-to-end distances specified by the equilibrium junction coordinates. If the crosslinks were formed in the system at equilibrium, then the total number of configurations for each strand of the pair is $\omega_1(g_1)_0$ and $\omega_2(g_2)_0$, and the number available to the pair is $\omega_1\omega_2(g_1)_0(g_2)_0$.

Suppose that $(f_i)_0$ is the fraction of the B pairs in topological class i from a total of Q such classes. The number of configurations available to a pair in class i will be $\omega_1\omega_2(g_1)_0(g_2)_0(f_i)_0$, and the total number of configurations in the undeformed state for B pairs is

$$\Omega_0 = \prod_{i=1}^{Q} [\omega_1\omega_2(g_1)_0(g_2)_0(f_i)_0]^{(f_i)_0 B}. \tag{7.27}$$

The pairs will all have the same new internal coordinates in the deformed state. Suppose that the fractions of configurations for free strands which have the specified end-to-end distances of the deformed state are g_1 and g_2, and that the fractions in the various classes for pairs formed from equilibrium with these coordinates are f_i, The number of configurations available to a pair in class i in the deformed state is therefore $\omega_1 \omega_2 g_1 g_2 f_i$. However, the number of strands in each class is always $(f_i)_0 B$, so the total number of configurations in the deformed state is:

$$\Omega = \prod_{i=1}^{Q} [\omega_1 \omega_2 g_1 g_2 f_i]^{(f_i)_0 B}. \tag{7.28}$$

The entropy change contributed by the B pairs is given by the Boltzmann equation:

$$\Delta S = k \ln \frac{\Omega}{\Omega_0} \tag{7.29}$$

which yields [using $\Sigma(f_i)_0 = \Sigma f_i = 1$]

$$\Delta S = kB \left[\ln \frac{g_1}{(g_1)_0} + \ln \frac{g_2}{(g_2)_0} + \sum_{i=1}^{Q} (f_i)_0 \ln \frac{f_i}{(f_i)_0} \right]. \tag{7.30}$$

The first two terms on the right of Eq. (7.30) will be recognized as the independent strand contribution to the entropy. The topological or entanglement contribution is then

$$\Delta S_e = kB \sum_{i=1}^{Q} (f_i)_0 \ln \frac{f_i}{(f_i)_0}. \tag{7.31}$$

The total pair-wise entanglement contribution is simply the sum of ΔS_e for strand pairs with all possible sets of internal coordinates. In a network of N strands there will be $N(N-1)/2$ such pairs. However, classification is only significant for pairs that are relatively close. Pairs separated by more than a few radii of gyration will belong to unentangled class exclusively. This assures that the total entropy change will be proportional only to N.

Equation (7.31) is a familiar form in statistical mechanics. For small displacements it is negative in sign and proportional to the square of displacement. For tensile deformation a power series expansion in the elongation ratio α, and with $\frac{\partial}{\partial \alpha} \Sigma f_i = 0$, yields

$$\Delta S_e = - \frac{kB(\alpha - 1)^2}{2} \sum_{i=1}^{Q} \frac{1}{(f_i)_0} \left(\frac{\partial f_i}{\partial \alpha} \right)_{\alpha=1}^{2} + 0((\alpha - 1)^3) \tag{7.32}$$

or, to terms of order $(\alpha - 1)^2$,

$$\Delta S_e = - \frac{k B (\alpha - 1)^2}{2} \sum_{i=1}^{Q} \left[f_i \left(\frac{\partial \ln f_i}{\partial \alpha} \right)^2 \right]_{\alpha = 1} \tag{7.33}$$

since all terms in the summation are positive, the topological contribution, ΔS_e, must be negative. Thus, in both the free energy of deformation $\Delta F_e = - T \Delta S_e$ and the tensile stress $f = \dfrac{1}{V_0} \dfrac{\partial \Delta F_e}{\partial \alpha}$, topological effects must always increase the initial modulus beyond the independent strand value. No general conclusions can be drawn about magnitude, nor about mechanical response at large deformation, other than to say that there is no reason to anticipate simple neo-Hookean behavior.

7.6.2. Topological Models for Entangling Interactions

Evaluating of the apportionment of configurations among topological classes is difficult even for simple models of non-crossing chains. Prager and Frisch (302) have analyzed the topological classes of an infinite rod and a closed-loop random coil. The random coil is allowed to pass freely through its own contour, so self-entanglement is not considered. The classes correspond simply to the number of loops the coil makes around the bar. Thus, Configurations A, B, and C

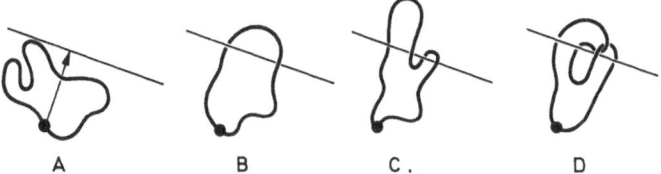

belong to different topological classes, while Configuration D is entangled with the bar only because of self-entanglement and therefore in this analysis is classified with configuration A.

The apportionment among classes depends on r, the distance of one specified point on the coil (the starting point of a self-closing random walk of n steps) from the bar. An exact solution is obtained for the probability $P_e(r)$ that the coil is entangled with the bar, i.e., that the coil is looped at least once around the bar. The average displacement of the coil produced by an external force F was then calculated from $P_e(r)$, using the Boltzmann equation. The force-displacement relation is non-linear, with an initial spring constant (initial slope of the force-displacement function) which is 1.66 times the value for a loop permanently fastened to the bar at one point. For large displacements, the spring softens and then again becomes linear in the force with approximately the same spring constant as an attached coil. This result is especially interesting because molecular springs which soften with deformation are exactly what is required to explain a positive C_2 term in the Mooney-Rivlin equation. The authors point out that a much simpler probability model gives a very similar force-displacement relation: the probability $P_0(r)$ that the starting point and midpoint of the coil lie on the opposite sides of the plane normal to r and containing the rod.

Edwards (300) examined the classification problem for a random coil bound at its end points and an infinite rod oriented normal to the end-to-end vector of the coil:

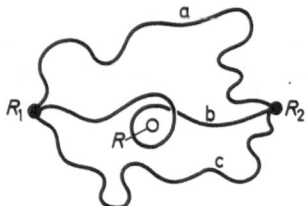

The ends of each coil (contour length L, step length 1) are attached to points R_1 and R_2 in a plane normal to the rod, which pierces the plane at R. The class of a configuration depends on whether the contour passes above the rod without encirclement (Configuration a), passes above with one turn around the bar (Configuration b), etc. An integer defining the class of any configuration is obtained from the angle swept out by the vector from R to a point on the path of the configuration as that point moves from R_1 to R_2. Thus,

$$\theta = R_1 R R_2 + 2\pi n \quad (-\infty < n < \infty) \tag{7.34}$$

and, for example, $n = 0$ from Configuration a, $n = -1$ for Configuration b, and $n = +1$ for Configuration c. Exact expressions for the fraction of configurations in each class were obtained, using an analogy in electromagnetic theory for the calculation of induced current in one conductor (the random coil) due to current flowing in a second, nearby conductor (the rod). Perhaps not unexpectedly, the actual structure of the computation turns out to be very similar to that of the restricted diffusion calculation used by Prager and Frisch. Alexander-Katz and Edwards (303) recently examined classification in a system comprising a random coil and an object of solenoidal configuration.

In general, it appears that the fraction of configurations in the various topological classes can be determined for models in which one of the elements is a fixed curve and the other is a random coil. The detailed calculations are intricate and difficult, however, and some simple generalizations are needed which could be used as a step towards building classification effects into the network theories. Classification for the case of two random coils and for self-entanglement are unsolved problems at the present time.

7.6.3. Network Models with Some Non-Classical Character

The results of these model calculations can be applied in a rather crude way to the network situation. One might consider each network strand individually, and in a rough way represent its topological interaction with each neighboring strand by a single probability $P_e(r_i)$ (from Prager and Frisch, for example), in which r_i is the mean distance between centers of gravity of the central strand and its ith neighbor. Thus, only two topological classes, entangled and not entangled, would be considered, with $(f_1)_0 = P_e$ and $(f_2)_0 = 1 - P_e$ for the interaction with each neighboring strand. The distances r_i and corresponding probabilities would change with deformation, and, if the deformation were affine, f_1 and f_2 for each neighbor could be calculated as a function of deformation. Assuming the total effect is obtainable by summing over individual strands (neglecting three-chain and higher order classification), the free energy and the modulus could be calculated.

Another approach to network modeling is suggested by the work of Edwards and Freed (304), Collins and Wragg (305), and Eichinger (306) on the configurations available to a confined chain. The idea is to represent the effects of environmental connectedness on the configurations of a network strand by a box. The configurations of a strand with junction points centered within the box could be classified according to whether or not the locus crosses the box surface at any point. Deformation might be assumed to alter the positions of both the junction points of the strand and the dimensions of the box in an affine manner. Classification in the deformed state could be evaluated, leading ultimately to an evaluation of ΔS_e by Eq. (7.31). Stress-strain behavior would presumably depend on the ratio $\langle r^2 \rangle_0 / L_0^2$.

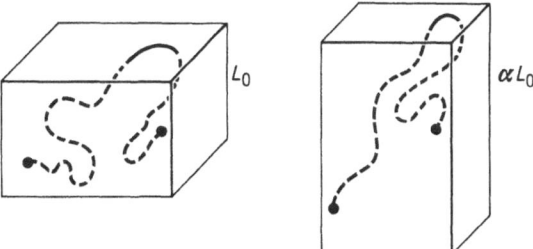

One could also construct a network involving elements which allow some adjustment in the location of their coupling points in response to deformation. Consider a tetrahedral network element (3) in which two gaussian strands of equal contour length are attached to the corners and joined together at their midpoints by a crosslink:

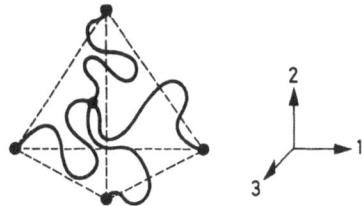

If the network is given a tensile deformation in the 2 direction and the corners move affinely, the four chains in the tetrahedron contribute independently to the modulus, with the well known result obtained from the rubber elasticity theory:

$$F_c = 4kT \frac{\langle r^2 \rangle}{\langle r^2 \rangle_0} \left(\alpha - \frac{1}{\alpha^2} \right) \tag{7.35}$$

in which F_c is the force, and the mean square end-to-end distances $\langle r^2 \rangle$ and $\langle r^2 \rangle_0$ refer to the four network strands in the unstressed network and as free chains respectively.

If the crosslink is removed, the system reverts to two independent chains. To preserve the symmetry of the system one needs to consider the average for three tetrahedra operating together, one for each of the three possible pairings of the strands. The result is again neo-Hookean behavior but with a lower modulus contribution.

$$F_{nc} = \frac{8}{3} kT \frac{\langle r^2 \rangle}{\langle r^2 \rangle_0} \left(\alpha - \frac{1}{\alpha^2} \right). \tag{7.36}$$

An intermediate situation is obtained if the central crosslink is replaced by a slipping link such that the two chains are constrained to be in contact at one

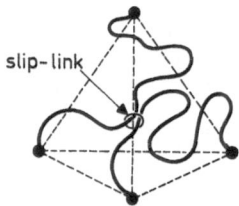

point, the contact point being allowed to lie anywhere along the chain contours. Thus, only configurations in which the two chains are in contact are counted in evaluating the change of entropy with deformation. Again, three tetrahedra are required to preserve symmetry. For this case, numerical solutions are required; some results (241) are summarized in Fig. 7.2 in the form of a Mooney-Rivlin plot. Briefly, one sees deformation softening with the largest effects obtained in super coiled systems ($\langle r^2 \rangle / \langle r^2 \rangle_0 < 1$). For swelled systems ($\langle r^2 \rangle / \langle r^2 \rangle_0 > 1$), the ratio $F \bigg/ \left(\alpha - \frac{1}{\alpha^2} \right)$ is nearly constant with a value which is practically the same as the system with no central link. In compression the slipping link in supercoiled systems actually contributes slightly more than a permanent cross-link to the modulus.

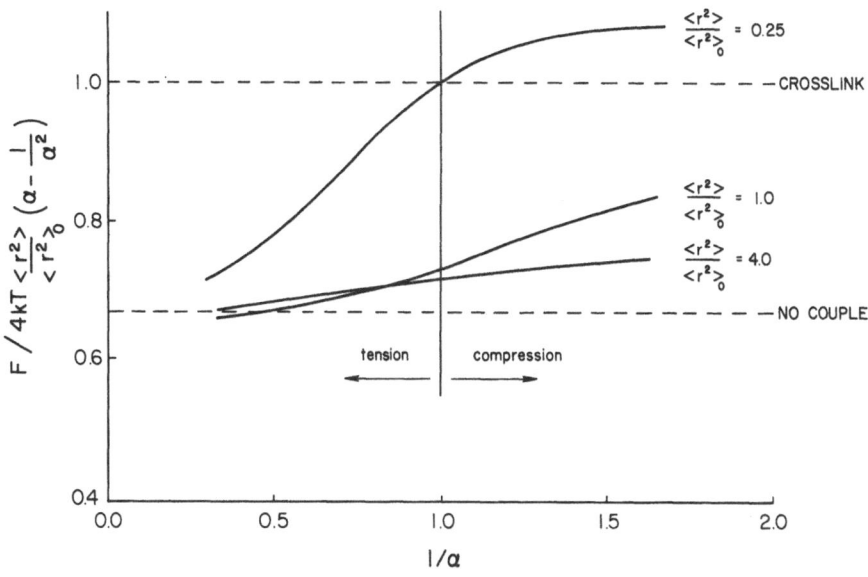

Fig. 7.2. Mooney-Rivlin plot for the Slip-Link model

Some of the effects displayed by the model are observed experimentally: $F / \left(\alpha - \dfrac{1}{\alpha^2} \right)$ is found to be a decreasing function of the extension ratio, and non-Hookean behavior is reduced by swelling. However, it is also reduced by supercoiling, which is not given by the model. Although probably not a good model of entanglement interaction, the slipping link at least illustrates that even simple configurational constraints can produce unexpected results in the stress-strain behavior. It also demonstrates that, as emphasized by Prager and Frisch, it is not configurational restriction *per se*, but, rather, changes in configurational restriction with deformation, which controls the modulus. Thus, swelling reduces the slip-link contribution, while intuition suggests the reverse. Finally, the result implies that entanglement (deduced from the plateau modulus in the system before crosslinking) and permanent crosslinks may not necessarily contribute in an additive fashion to equilibrium mechanical properties, even in the region of small deformations.

8. Non-linear Viscoelastic Properties

8.1. Introduction

Only a few non-linear viscoelastic properties have been studied with polymers of well-characterized structure. The most prominent of these is the shear-rate dependence of viscosity. Considerable data have now been accumulated for several polymers, extending over a wide range of molecular weights and concentrations. Several possible mechanisms for the shear-rate dependence of viscosity have been suggested, some of them based on the supposition of flow-induced changes in the entanglement density. One purpose of this section is to review the data on $\eta(\dot\gamma)$ in order to evaluate these theories. The steady state normal stress functions N_1 and N_2 are also non-linear properties. A limited amount of data is available on $\Psi_1(\dot\gamma)$. The coefficient at low shear rates, $\Psi_1(0)$, has already been discussed in Section 5 because of its theoretical relationship to linear viscoelastic parameters. Data on $\Psi_2(\dot\gamma)$ are still scanty, which is also the case for time-dependent flow properties such as the growth and relaxation of stress at the onset and cessation of steady shearing flow. A brief review of these latter properties concludes with a discussion of the possible role of entanglements in non-linear viscoelasticity generally. As in the examination of linear viscoelastic properties, emphasis is placed on systems of narrow molecular weight distribution, with only brief comments on the effects of polydispersity.

8.2. Shear Rate Dependence of Viscosity

The viscosity approaches a constant value η_0 at low shear rates; experimentally this region is reached in narrow distribution polymers without difficulty. As the shear rate is increased the viscosity departs from η_0, becoming in almost all cases a decreasing function of shear rate. In many of these it attains an approximate power-law dependence at high shear rates: $\eta \propto |\dot\gamma|^{-d}$. In concentrated systems the transition from η_0 to power-law behavior takes place over a relatively narrow range of shear rates. Polydispersity, particularly the presence of a high molecular weight tail, widens the transition and displaces the departure from η_0 to lower shear rates. In systems of very broad distribution, such as commercial polyethylene, the shear rates required to reach $\eta_0(\dot\gamma \approx 10^{-4} - 10^{-5}\,\text{sec}^{-1})$ may be very difficult to reach in practice. At very high shear rates the viscosity probably levels off again. However, except in dilute solution (307) or for low molecular weight polymers (308), this range is seldom attainable because of mechanical degradation or the appearance of various types of flow instability.

The onset of shear rate dependence can be specified by a characteristic shear rate $\dot\gamma_0$, chosen for example as the shear rate at which η has fallen to some arbitrary fraction of η_0. Experimentally $\dot\gamma_0$ is almost identical with ω_0, its counterpart in the frequency dependence of the dynamic viscosity $\eta'(\omega)$ (170, 309). *Review of the current literature reveals no polymer system which violates this principle.* Since the departure of η' from η_0 is dictated by the longest relaxation times in the linear viscoelastic spectrum, the onset of shear rate dependence must be similarly controlled. This is not to say however that $\eta(\dot\gamma)$ and $\eta'(\omega)$ are mechanistically related [early suggestions along these lines were made by Bueche (310) and Pao (311)], although their forms are sometimes similar or even partially superimposeable by slight adjustments in the $\dot\gamma$ or ω scales. Flow birefringence shows that chain configuration is affected quite differently in the two types of deformation (307), so the resemblance can be no more than a superficial one. Also, the forms of the two functions are in fact different, $\eta'(\omega)$ decreasing more rapidly than $\eta(\dot\gamma)$ and sometimes reaching a slope,

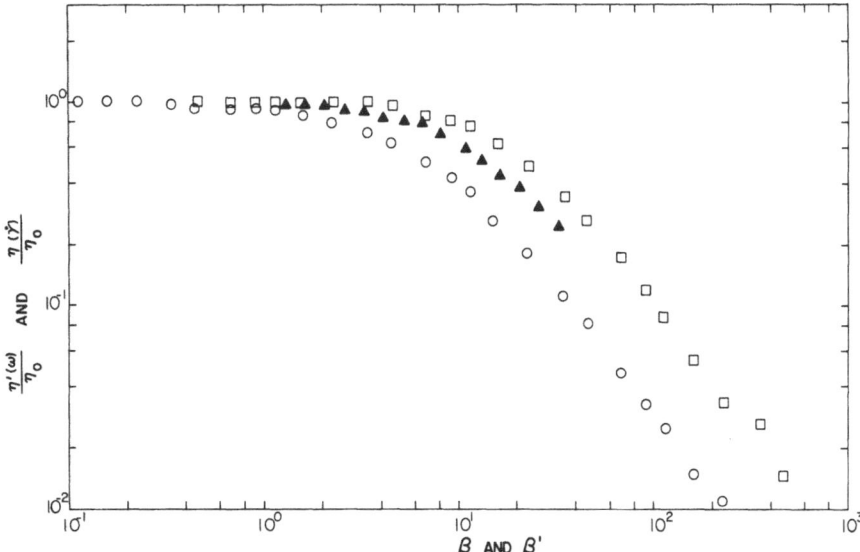

Fig. 8.1. Dynamic viscosity $\eta'(\omega)$ and steady state viscosity $\eta(\dot\gamma)$ for undiluted narrow distribution polystyrenes. The data are plotted in reduced form to facilitate comparison. The dimensionless shear rate or frequency is $\eta_0 \bar{M}_w \dot\gamma / \varrho RT$ or $\eta_0 \bar{M}_w \omega / \varrho RT$. [See Eq.(8.3)]. The dynamic viscosities are for $\bar{M}_w = 215000$ ($\bigcirc$) and $\bar{M}_w = 581000$ ($\square$) at 160° C (312). The steady shear viscosity is for $\bar{M}_w = 411000$ ($\blacktriangle$) at 176° C (313). The shapes in the onset region are similar for the three curves, but the apparent limiting slope for the dynamic viscosities is about -1.3

$-d(\log \eta')/d(\log \omega)$, greater than unity in narrow distribution systems (Fig. 8.1). Since the shear stress must be a monotonically increasing function of shear rate, the slope, $-d(\log \eta)/d(\log \dot\gamma)$, can never exceed unity for real flows, and experimentally it seems to be no greater than about 0.85.

The absolute value of the complex viscosity,

$$|\eta^*(\omega)| = [G'^2 + G''^2]^{1/2}/\omega \tag{8.1}$$

has a slope which can never exceed unity, and often $|\eta^*(\omega)|$ agrees rather closely with the form of $\eta(\dot\gamma)$ (314) (Fig. 8.2). The data in Fig. 8.3 show that the correspondence is not a general one however. The objections to a fundamental connection between η' and η must also apply to the case of $|\eta^*|$ and η.

The Rouse and Zimm models provide little direct help in dealing with $\eta(\dot\gamma)$ since each predicts a viscosity which is independent of shear rate. The principal interest here is in concentrated systems where entanglement effects are prominent. Nevertheless, shear rate can influence the viscosity of polymer systems at all levels of concentration, including infinite dilution (307) and melts with $M < M_c$ (308, 315). It is therefore essential to identify the causes of shear rate dependence in systems of isolated or weakly interactions molecules in order to separate intramole-

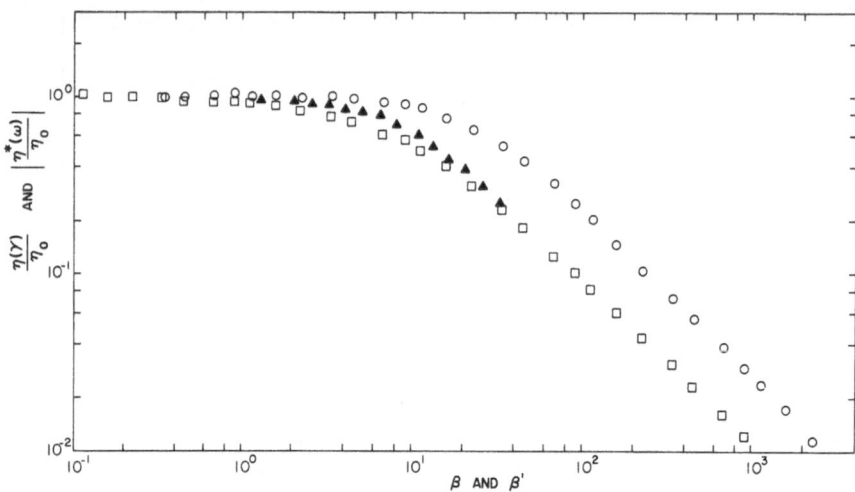

Fig. 8.2. Absolute complex viscosity $|\eta^*(\omega)|$ and steady state viscosity $\eta(\dot{\gamma})$ for undiluted narrow distribution polystyrenes. Data are shown for the same samples as in Fig. 8.1. Limiting slopes for both $\eta(\dot{\gamma})$ and $|\eta^*(\omega)|$ are almost identical at ca. -0.8

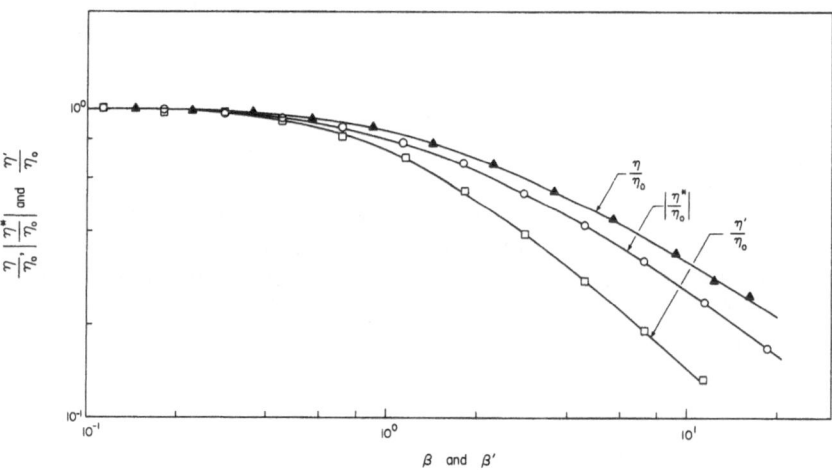

Fig. 8.3. Dynamic viscosity, absolute complex viscosity, and steady state viscosity for narrow distribution polystyrene. Data obtained at $25°$ C on a 0.071 gm/ml solution of polystyrene ($\bar{M}_w = 860000$) in Aroclor (*316*)

cular and intermolecular contributions at high concentration and molecular weight.

8.2.1. Correlation of Experimental $\eta(\dot{\gamma})$ Behavior

The effects of molecular structure on $\eta(\dot{\gamma})$ are most conveniently discussed in terms of reduced variables. The choice of form for reduced shear rate is guided both by the observation that the long relaxation times govern the onset of shear

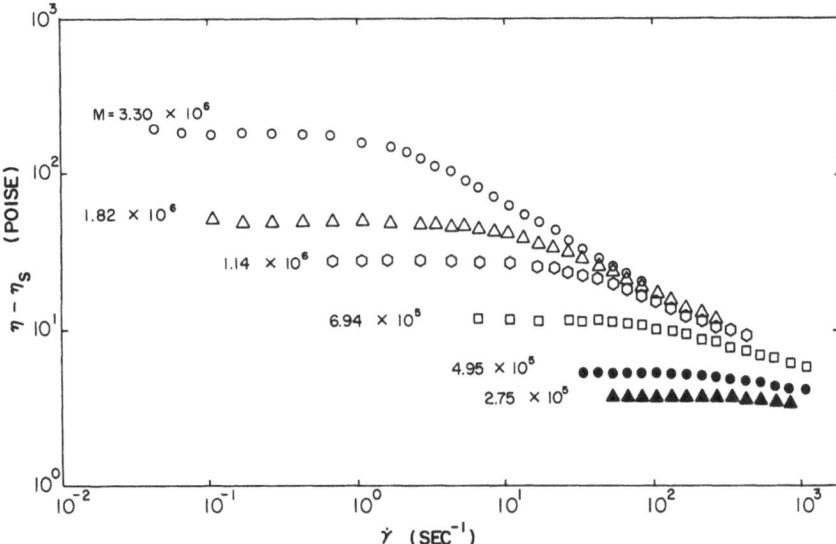

Fig. 8.4. Shear rate dependence of viscosity as a function of molecular weight at low concentrations. Data were obtained at 25° C on narrow distribution poly(α-methyl styrene) samples at a concentration of 0.02 gm/ml in kanechlor (*198*)

rate dependence and by the form of the characteristic relaxation time in spring-bead models:

$$\frac{\eta - \eta_s}{\eta_0 - \eta_s} = f(\beta),\qquad(8.2)$$

and

$$\beta = (\eta_0 - \eta_s)\, M\dot{\gamma}/cRT.\qquad(8.3)$$

The reduced shear rate β retains a meaning over the entire range of concentration, becoming $\beta = [\eta]\,\eta_s M\dot{\gamma}/RT$ at infinite dilution and $\beta = \eta_0 M\dot{\gamma}/cRT$ in concentrated solutions, where $\eta_0 \gg \eta_s$. The corresponding reduced viscosities are $[\eta]/[\eta]_0$ and η/η_0[7]. Reduced plots remove most of the observed variation among systems (Figs. 8.4–8.7), and simplify the empirical examination of residual

---

[7] Others have defined these variables slightly differently, using $(1-\varphi)\eta_s$ in place of η_s in Eqs.(8.2) and (8.3), where φ is the volume fraction of polymer in the solution. However, the practical effect is entirely negligible for all data examined in this review. The value of $\eta_0 - \eta_s$ increases so rapidly with polymer concentration that $\eta_0 - \eta_s$ and $\eta_0 - (1-\varphi)\eta_s$ are essentially indistinguishable at all levels of concentration.

Another alternative is to correlate data in terms of a reduced shear stress, $\dfrac{\sigma' M}{cRT}$, in which σ' is $(\eta - \eta_s)\dot{\gamma}$, or simply $\eta\dot{\gamma} = \sigma$, when $\eta \gg \eta_s$. The practical advantage of shear stress as a correlating variable is that it emphasizes the sharpness of the down-turn in reduced viscosity beyond the onset of non-Newtonian behavior. Also, the magnitudes of other non-Newtonian properties, such as the first normal stress difference, appear to be controlled by the shear stress magnitude. For example, N_1 vs σ is practically independent of temperature for a given sample, while N_1 vs $\dot{\gamma}$ is very temperature sensitive. For our purposes here the choice is of course arbitrary.

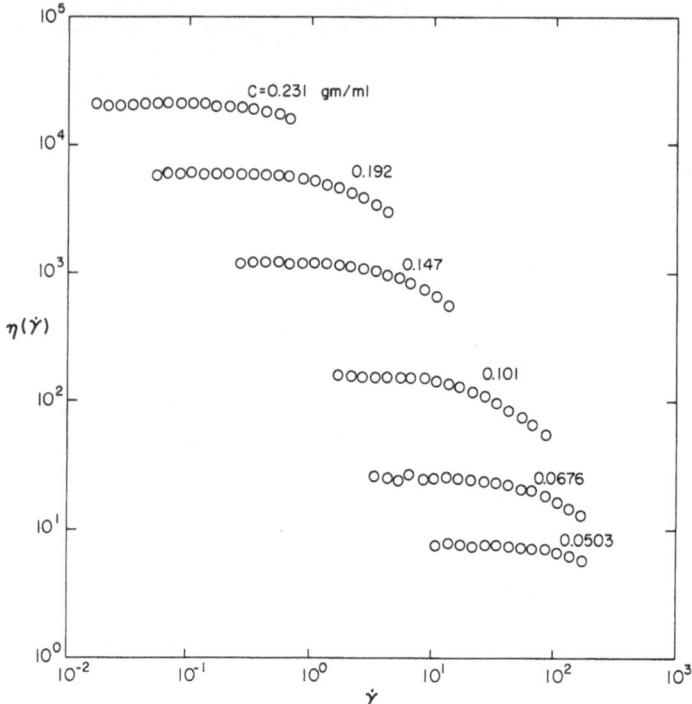

Fig. 8.5. Shear rate dependence of viscosity as a function of concentration. Data were obtained at 30° and 50° C on one narrow distribution sample of poly(α-methyl styrene) $\bar{M}_w = 1\,820\,000$) in α-chloronaphthalene (199)

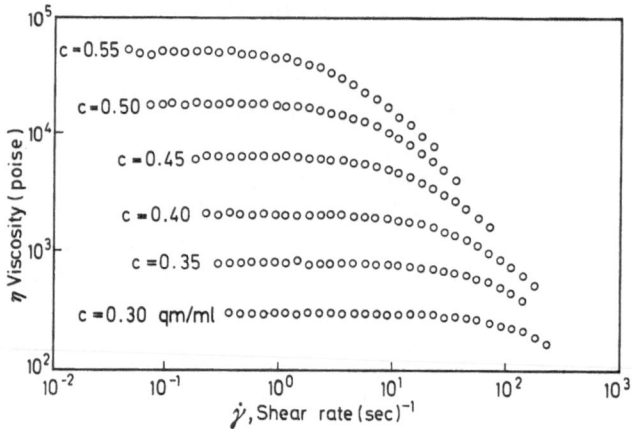

Fig. 8.6. Shear rate dependence of viscosity as a function of concentration. Data were obtained on a single narrow distribution sample of polystyrene ($\bar{M}_w = 411\,000$) in n-butyl benzene (155) at 30° C. (Reproduced from Transactions of the Society of Rheologie, Volume 11, Fig. 2, p. 273, New York: Wiley & Sons.)

differences related to structure (Fig. 8.8). These differences are of two kinds: the form of the reduced viscosity function $f(\beta)$, and the value of reduced shear rate β_0 which locates the onset of shear rate dependence. *For the purposes of this review the form of $f(\beta)$ is characterized by an estimated value of the power-law*

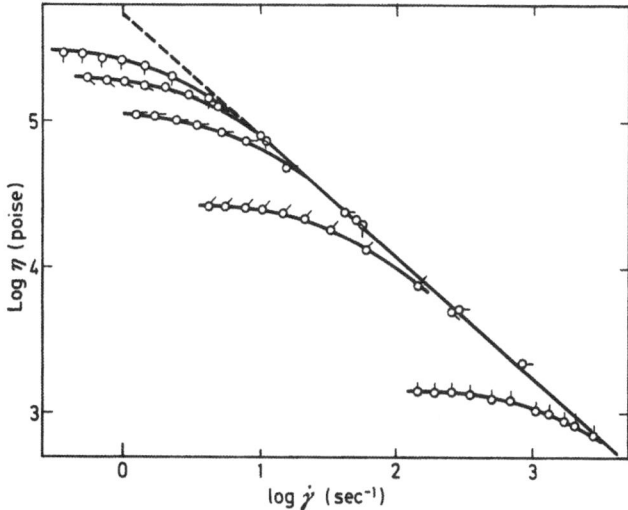

Fig. 8.7. Shear rate dependence of viscosity for undiluted polystyrenes of different molecular weights. Weight-average molecular weights are ⊘ 48 500; ⊘ 117 000; ⊖ 179 000; ⊘ 217 000; ⊙ 242 000. All data are reduced to 183° C; the dashed line has a slope of −0.82 (324). (Reproduced from Journal of colloid and interfache science, Vol. 22, Fig. 1, p. 520. New York: Academic Press.)

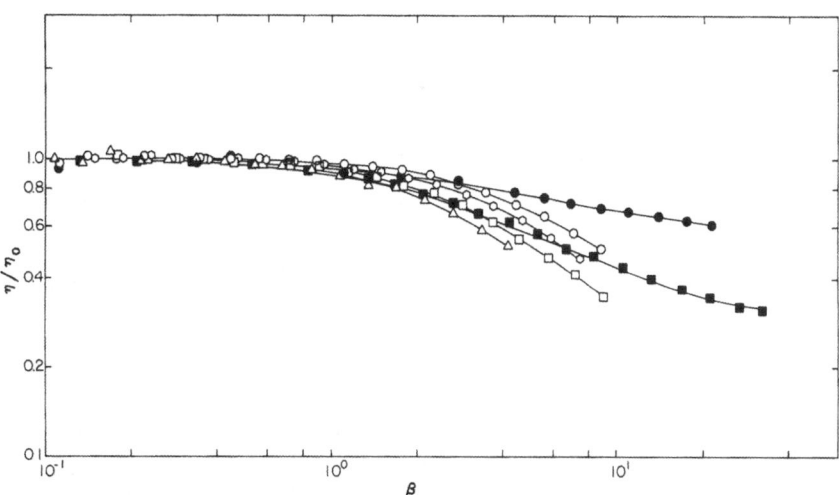

Fig. 8.8. Shear rate dependence of viscosity in reduced form as a function of concentration. Data were obtained on one narrow distribution sample ($\bar{M}_w = 1\,820\,000$) of poly(α-methyl styrene) in two solvents (198, 199). Symbols are: ● for 0.00552 gm/ml in α-chloronaphthalene (CN), ■ for 0.0231 in CN, △ for 0.0676 in kanechlor (K), □ for 0.101 in K, ◇ for 0.147 in K, and ○ for 0.192 in K. Note the progressive increase in slope with concentration at low concentrations, followed by a progressive shift to large β values without much shape change at high concentrations

exponent d. The critical shear rate β_0 is chosen as the value of reduced shear rate at which $\eta - \eta_s$ falls to 80% of its value at zero shear rate. There is no special significance to these definitions. Much information on the detailed shape of $f(\beta)$ is ignored of course, and other definitions of β_0 would also suffice. Each

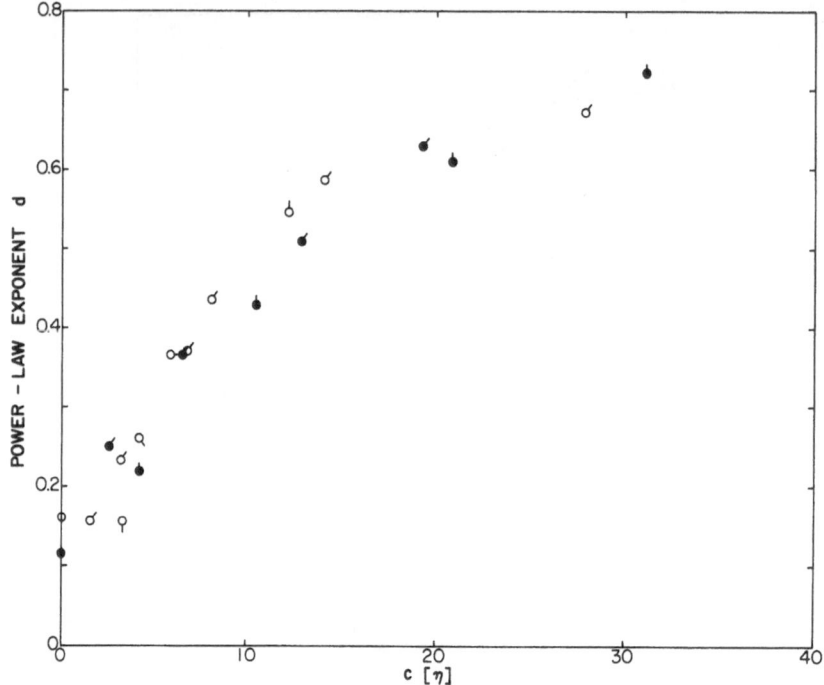

Fig. 8.9. Power law exponent d as a function of the coil overlap parameter $c[\eta]$ at low concentrations. The filled circles are narrow distribution polystyrene solutions (*177, 316, 318*), the open circles are poly(α-methyl styrene) (*198, 318*). Solvents are chlorinated di-phenyls except the intrinsic viscosity data which were obtained in toluene. Symbols are: for polystyrene ● $M = 13.6 \times 10^6$, ◖ 1.8×10^6, and ◗ 0.86×10^6; for poly(α-methyl styrene) ○ $M = 7.5 \times 10^6$, ◔ 3.3×10^6, ◑ 1.82×10^6, ◐ 1.14×10^6, ◓ 0.694×10^6, and ♀ 0.444×10^6

parameter can be evaluated with reasonable accuracy from the available data and can be examined without reference to any particular molecular viewpoint.

8.2.1.1. Form of the Reduced Viscosity Function. Endo *et al.* have published extensive data on $\eta(\dot{\gamma})$ in 1–10% solutions of poly(α-methyl styrene) (*198*). Ashare (*177*) and Harris (*316*) have covered a similar range of concentrations and molecular weights for polystyrene. Data on the shear rate dependence of intrinsic viscosity for the two polymers are also available (*318–321*). It is apparent from Fig. 8.8 that the viscosity-shear rate curves are not superimposable at these moderate concentrations, $f(\beta)$ changing with both concentration and molecular weight throughout the regime[8]. The shape, as characterized by the power-law

---

[8] It is conceivable that a suitable reducing scheme could be found to unify the $\eta - \dot{\gamma}$ at intermediate concentrations. For example, the viscosities in Fig. 8.4 tend to level off at high shear rates. Aside from the two lowest molecular weights, the limiting viscosity $\eta_\infty - \eta_s$ is approximately the same for all. The same data were replotted in the recently suggested form (*317*), $(\eta - \eta_\infty)/(\eta_0 - \eta_\infty)$ vs β, in hopes of obtaining a master curve. A complete reduction was not obtained, although this same method was successfully used with similar data on a series of undiluted poly(dimethyl siloxane) samples (*317*). Possibly a reduced form based on $\beta' = \beta(\eta_0 - \eta_\infty)/(\eta_0 - \eta_s)$ or some other variant would have been more successful.

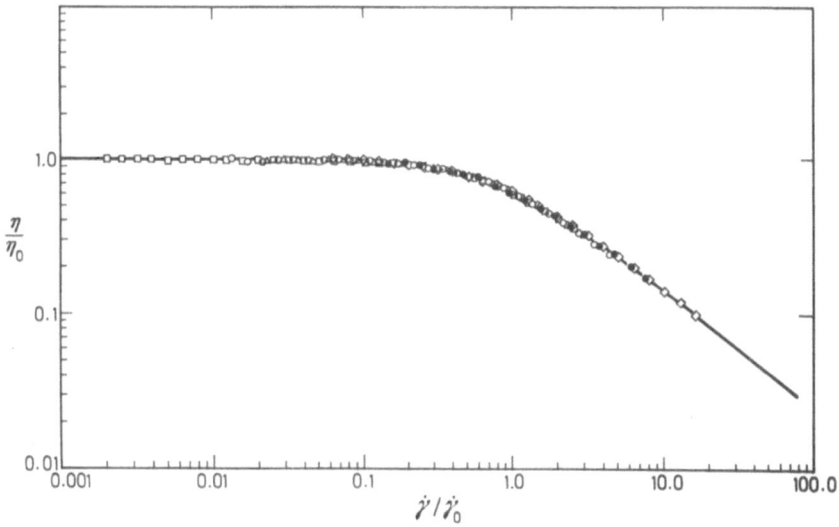

Fig. 8.10. Viscosity-shear rate master curve for concentrated polystyrene-n-butyl benzene solutions. The data were obtained for molecular weights ranging from 160000 to 2400000; concentrations from 0.255 to 0.55 gm/ml, and temperatures from 30° C to 60° C (155)

exponent, appears to be principally a function of the coil overlap parameter $c[\eta]$. Figure 8.9 shows d vs $c[\eta]$. The exponent rises from $d \approx 0.10$ at infinite dilution and approaches $d \approx 0.8$ for $c[\eta] > 20$.

Although taking place over a somewhat wider range of $c[\eta]$, this transition parallels the variation of J_{eR} from Zimm-like to Rouse-like behavior at low concentrations (15). It also supports the contention (Section 5) that coil overlap is the principal structural variable affecting viscoelastic behavior at low to moderate concentrations.

Some variations in $f(\beta)$ with solvent power and molecular weight are observed in the intrinsic viscosity (318–321). Significant shear rate dependence in $[\eta]$ at convenient shear rates seems to require very high molecular weights, and is enhanced in good solvents (318). The data available at low to moderate concentrations were obtained only in good solvents, so the variation with solvent power in this range is not known. The agreement between data for polystyrene and poly(α-methyl styrene) in Fig. 8.9 suggests the possibility of a universal correlation between the form of $f(\beta)$ and $c[\eta]$ for linear polymers of narrow molecular weight distribution. However, these two polymers are probably not sufficiently different to justify such a broad conclusion at this point. Also, there is some indication in the data at low concentrations that $(\eta - \eta_s)/(\eta_0 - \eta_s)$ approaches a finite limit at high shear rates (Fig. 8.4). The source of this behavior could be specific to each polymer although, as in the case of $\eta'(\omega)$ where internal viscosity provides a lower limit at high frequencies, it probably has little influence on $f(\beta)$ except at very high shear rates.

Judged by the superposability of viscosity-shear rate data on the same master curve for a variety of polymers [polystyrene (155) (Fig. 8.10), poly(α-methyl

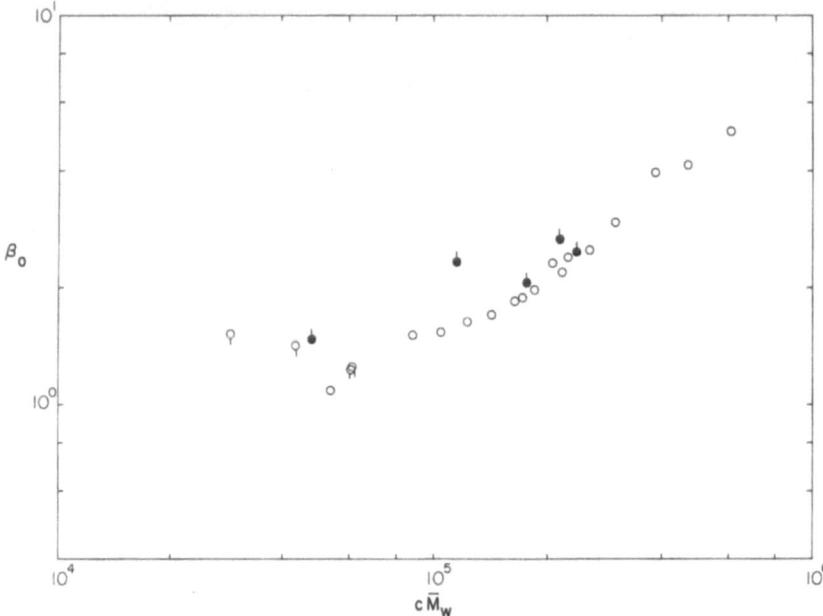

Fig. 8.11. Dimensionless shear rate β_0 locating the onset of shear rate dependence for viscosity in narrow distribution polystyrene systems. Symbols are: ○ for solutions at 30° C in n-butyl benzene (*155*), ⵕ for solutions at 25° C in arochlor (*177*), and ● for undiluted polymers at 159° and 183° C (*324*). Values for the intrinsic viscosity ($cM = 0$) lie in the range $\beta_0 = 1–2$, varying somewhat with solvent-polymer interaction and molecular weight (*307, 318–321*)

styrene) (*199*), polyvinyl acetate (*195*), polybutadiene (*322*), and polydimethyl siloxane (*323*)], *the form of $f(\beta)$ in concentrated solutions and melts appears to become essentially a universal function for monodisperse linear chains of high molecular weight.* Based on Fig. 8.9 this limiting form applies for systems in which $c[\eta] \gtrsim 30$.

8.2.1.2. Onset of Shear Rate Dependence. At low concentrations β_0 is nearly constant at 1.5–2.0, values not much different than those at infinite dilution. At high concentrations and molecular weights the β_0 values increase systematically. As in the earlier correlations of steady state compliance, the departure appears to be a function of the segment-segment interaction parameter cM. Figure 8.11 shows the behavior in polystyrene systems, including results for melts (*324*) and for solutions in various solvents (*155, 177, 316*). Other studies have shown that solvent power has very little influence on β_0 (*53, 325*). The departure from constant β_0 begins somewhat beyond $cM = \varrho M_c$, and corresponds roughly to the region where $f(\beta)$ ceases to vary with c and M. Figure 8.12 is a similar plot for poly(α-methyl styrene). The apparent trend to larger values of β_0 at small cM is believed to be an artifact of the definition of β_0 used here and the flatness of $f(\beta)$ for dilute solutions of all but the very highest molecular weight samples. Values of β_0 at zero concentration reported with the figure were obtained from $[\eta]$ vs $\dot{\gamma}$ data on poly(α-methyl styrene) of much higher molecular weight

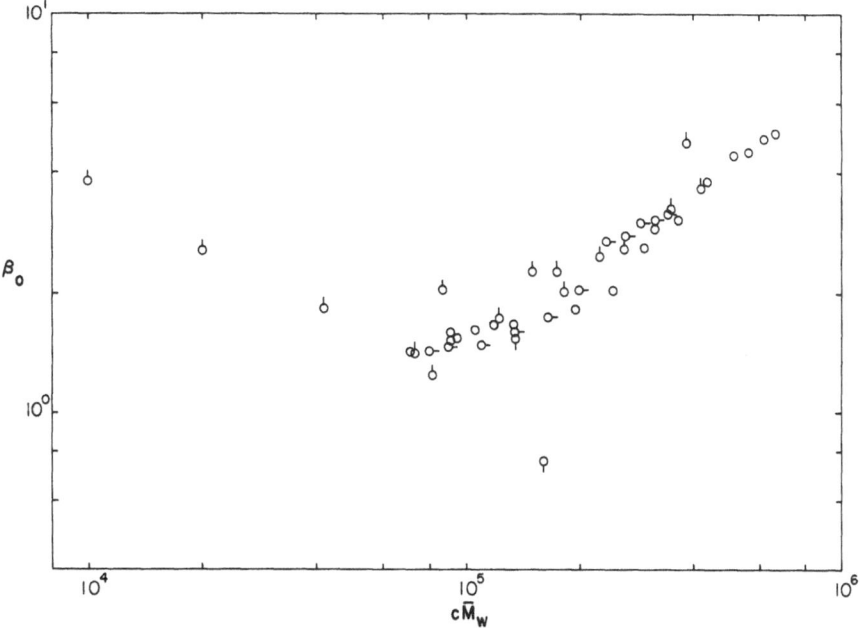

Fig. 8.12. Dimensionless shear rate β_0 locating the onset of shear dependence in the viscosity for narrow distribution poly(α-methyl styrene) systems. Symbols are (198, 199): ○ $M = 3.3 \times 10^6$, ◌ $M = 1.82 \times 10^6$, ○- $M = 1.19 \times 10^6$, and ○ $M = 0.444 \times 10^6$. Values for intrinsic viscosity ($cM = 0$) are similar to those for polystyrene (see caption of Fig. 8.11)

($M = 7 \times 10^6$) than the sample ($M = 1.8 \times 10^6$) which shows the apparent rise in β_0 in dilute solution.

Following the procedure used with J_{eR} (Section 5), β_0 data on these and other polymers were correlated in terms of $cM/\varrho M_c$, with results which are shown in Fig. 8.13. The parallel in behavior between the β_0 and J_{eR} master correlations is unmistakable. Even the relative positions of polymers on the master correlations are similar; note for example the data on J_{eR} and β_0 for polybutadiene. Published data on relatively narrow distribution polyethylene (210, 326) have not been included in Fig. 8.13 because departures from η_0 were too small to define β_0 with accuracy. However, estimates of β_0 from the data provided suggest that polyethylene may follow a different pattern than other polymers. Departures from η_0 seem to appear at anomalously low shear rates (326). Aside from η_0 values, viscoelastic data on well characterized crystallizable polymers in the melt state are rather scarce. Although not especially anticipated, it is certainly conceivable that crystallizability confers some unusual features to the flow behavior.

The parallel between β_0 and J_{eR} has been noted elsewhere (208, 213, 328, 329), and is not in fact fortuitous. It follows rather directly from the empirical observation that departures of $\eta(\dot{\gamma})$ and $\eta'(\omega)$ from η_0 are governed by the longest relaxation times of the system, combined with slight extensions of a reduced variables argument suggested by Markovitz for linear viscoelastic behavior (329). Suppose one wants to compare the forms of the dynamic moduli on

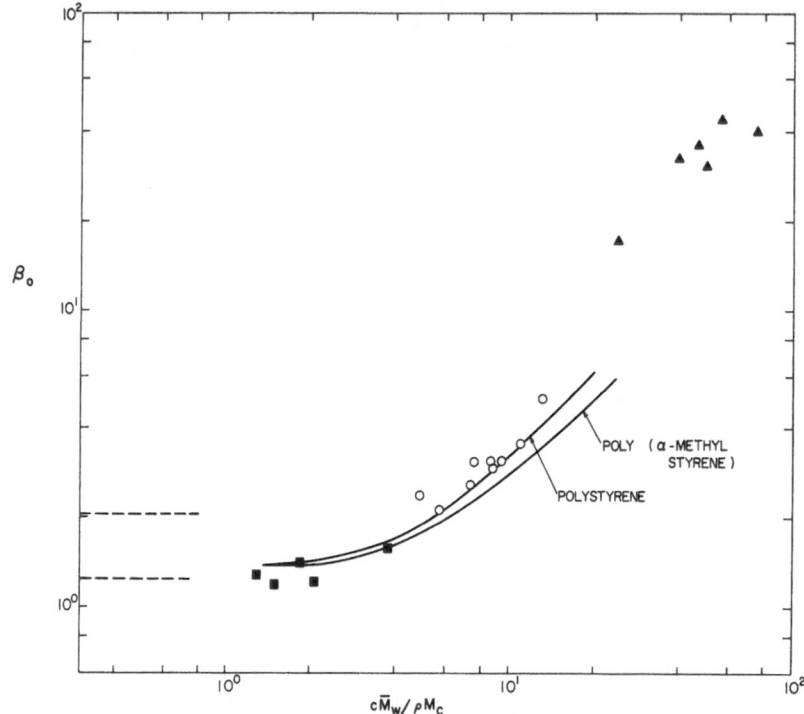

Fig. 8.13. Dimensionless shear rate β_0 locating the onset of shear rate dependence in the viscosity in narrow distribution systems of linear polymers vs $cM/\varrho M$. Symbols for data on additional polymers are: ▲ for undiluted 1,4 polybutadiene (*322*), ■ for undiluted poly(dimethyl siloxane) (*323*), and ○ for solutions of polyvinyl acetate in diethyl phthalate (*195*). The dotted lines indicate the ranges of β_0 for the intrinsic viscosity

samples having different values of η_0 and J_e^0. If solvent is present and $G'(\omega)$ and $G''(\omega) - \omega\eta_s$ are the polymer contributions, then as usual

$$\lim_{\omega \to 0} \frac{G''(\omega) - \omega\eta_s}{\omega} = \eta_0 - \eta_s, \tag{8.4}$$

$$\lim_{\omega \to 0} \frac{G'(\omega)}{\omega^2} = \eta_0^2 J_e^0. \tag{8.5}$$

If the dynamic moduli are expressed in the reduced forms,

$$G_R' = J_e^0 \left(\frac{\eta_0}{\eta_0 - \eta_s} \right)^2 G', \tag{8.6}$$

$$G_R'' = J_e^0 \left(\frac{\eta_0}{\eta_0 - \eta_s} \right)^2 (G'' - \omega\eta_s), \tag{8.7}$$

$$\omega_R = \frac{\eta_0^2 J_e^0}{\eta_0 - \eta_s} \omega \tag{8.8}$$

it is then easy to show that

$$\lim_{\omega_R \to 0} \frac{G_R''}{\omega_R} = 1 \,, \tag{8.9}$$

$$\lim_{\omega_R \to 0} \frac{G_R'}{\omega_R^2} = 1 \,. \tag{8.10}$$

Thus, differences among samples due to differences in J_e^0 and η_0 are removed in this reduced form. Since η_0 and J_e^0 are sensitive characteristics of the long relaxation time processes, and since the initial departure of $\eta_0' - \eta_s$ from $\eta_0 - \eta_s$ is governed by these same processes, one would espect significant departures from η_0 to occur near some roughly constant characteristic value of ω_R with a value of approximately unity. Finally, if the onset of shear rate dependence depends on the same processes, then Eq.(8.8) also gives an appropriate reduced form for $\dot{\gamma}$:

$$\dot{\gamma}_R = \frac{\eta_0^2 J_e^0}{\eta_0 - \eta_s} \dot{\gamma} \,. \tag{8.11}$$

With the definitions of β and J_{eR}, Eq.(8.11) can be expressed:

$$\dot{\gamma}_R = J_{eR}\beta \,. \tag{8.12}$$

The onset of shear rate dependence therefore is governed by the product $J_{eR}\beta$ and should occur near $J_{eR}\beta = 1$. The observed value of this product at departure will of course depend on the criterion chosen to define departure. It may also vary somewhat if the criterion requires a large departure and the form of $\eta(\dot{\gamma})$ differs substantially from sample to sample.

Figure 8.14 shows the result of combining the correlations of β_0 (Fig. 8.13) and J_{eR} (Fig. 5.18). The product $\beta_0 J_{eR}$ is remarkably independent of concentration and molecular weight from infinite dilution all the way to the undiluted melt state and shows no substantial variation from one polymer to another. For the particular definition of critical reduced shear rate $\dot{\gamma}_0$ used here, the experimental result for narrow distribution linear chains can be expressed as

$$\beta_0 J_{eR} = \frac{\eta_0^2 J_e^0 \dot{\gamma}_0}{\eta_0 - \eta_s} = 0.6 \pm 0.2 \,. \tag{8.13}$$

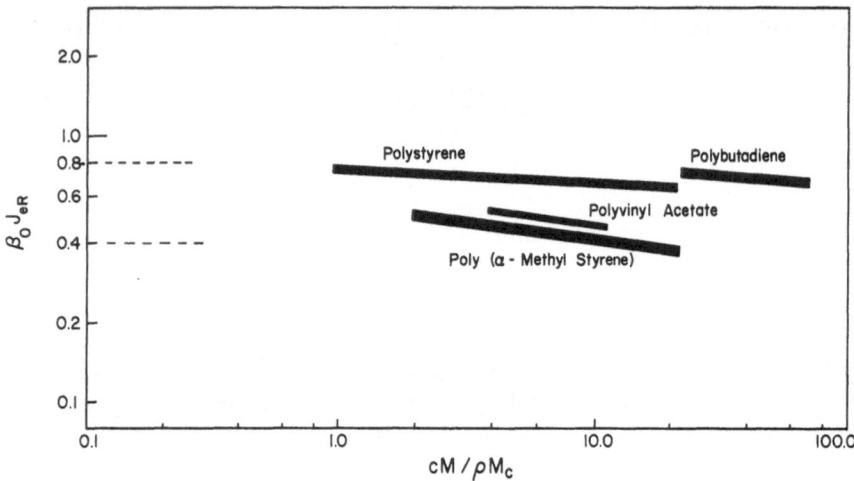

Fig. 8.14. Product of reduced compliance J_{eR} and characteristic shear rate β_0 for narrow distribution systems of linear polymers. The dashed lines indicate the range of $\beta_0 J_{eR}$ for the intrinsic viscosity

Aside from minor differences this equation is the same as that suggested recently by Prest et al. (208)[9].

There is considerable evidence to suggest that such correlations can be extended to polydisperse linear polymers. The fact that J_e^0 tends to be unusually large in branched polymers and that $\eta_0 \dot\gamma_0$ is unusually small suggests that the correlation may be applicable in branched systems also. If so, the result should have considerable empirical utility. It must be pointed out, however, that the definition of $\dot\gamma_0$ used here is a rather arbitrary one, and that the result in any case, like the Merz-Cox rule, has no foundation in fundamental continuum theories. For example, although the implied relationship between elastic properties and onset of shear rate dependence in the viscosity is acceptable within the framework of simple fluid theory, it is not a direct and unique consequence (330). This is not surprising, of course, since both the generalized Newtonian liquid and the Lodge elastic liquid are simple fluids, and yet the former has a shear rate dependent viscosity without elasticity while the latter has elasticity but a viscosity which is independent of shear rate.

---

[9] This correlation brings together two definitions of characteristic time from continuum rheology. Slattery (329a) notes that Bird has suggested a characteristic time τ_B for a viscoelastic fluid, defined in terms of the onset of shear rate dependence of the steady shear viscosity. For the correlations here we may take $\tau_B = 1/\dot\gamma_0$. He also notes that Truesdell had suggested a characteristic time τ_T, defined in terms of the normal stress differences in steady shear flow. For the correlations here we may take $\tau_T = \Psi_1(0)/\eta_0$. From the Coleman-Markovitz relation [Eq. (3.21)] $\tau_T = 2 J_e^0 \eta_0$, yielding, from Eq. (8.13):

$$\tau_T/\tau_B = (1.2 \pm 0.4) \, \frac{\eta_0 - \eta_s}{\eta_0} \qquad (8.13\,a)$$

for the special case of narrow distribution polymers.

8.2.2. Theories of Shear Rate Dependence of Viscosity

8.2.2.1. Intramolecular Contributions. Several explanations have been suggested for shear rate dependence of viscosity in dilute solutions. Shearing flow tends to extend the molecules in the flow direction, increasing the pervaded volume and reducing the average concentration of segments within the coil (*331*). Perturbation of coil dimensions due to the effects of excluded volume is therefore smaller in flow than at equilibrium. On the other hand, the surface of the pervaded volume is greater in the flow-extended state, so more segments are exposed to the drag forces of the external flow. The original Zimm analysis accounts for neither effect. According to Peterlin (*332*), the two effects generally combine to produce a net decrease in viscosity with increasing shear rate, beginning near $\beta = 1$. However, under some conditions, such as in the case of highly viscous good solvents, the viscosity may actually increase with shear rate (*333*).

Although shear rate effects are more pronounced in good solvents, the intrinsic viscosity decreases with shear rate even in θ-solvents, where excluded volume is zero (*317, 318*). The Zimm model employs the hydrodynamic interaction coefficients in the mean equilibrium configuration for all shear rates, despite the fact that the mean segment spacings change with coil deformation. Fixman has allowed the interaction matrix to vary in an appropriate way with coil deformation (*334*). The initial departure from $[\eta]_0$ was calculated by a perturbation scheme, and a decrease with increasing shear rate in θ-systems was predicted to take place in the vicinity of $\beta = 1$.

In the above calculations the sub-molecule springs of the model are linear, so the molecules are treated as though they were indefinitely extensible. Changes in viscosity due to alterations in excluded volume or hydrodynamic interaction are assumed to occur well before the mechanical non-linearity of real molecules at high extensions becomes important. Light scattering measurements during shearing flow show that the coils are indeed extended in the flow direction, although perhaps not as much as predicted by the linear spring-bead models (*335–337*). It turns out that finite extensibility can also provide a shear rate dependent viscosity.

There is nothing to prevent one from formulating diffusion equations for spring-bead models with non-linear springs. The difficulty comes in the calculation of stresses from these equations. The many-bead diffusion equation is very delicate. Seemingly simple modifications of the linear spring-bead model can prevent the uncoupling of motions by transformation to normal coordinates, thereby rendering the equations intractible. Inferences must then be drawn from either approximate treatments of many element models or exact treatments of low-order homologs for which the calculations can be carried to completion without a separation of variables. The low-order homolog of the linear spring-bead models is the elastic dumbbell: two frictional beads connected by a linear spring. Except for the distribution of relaxation times the elastic dumbbell has the same qualitative features, including indefinite extensibility, as the multiple bead Rouse model.

Bird and co-workers have recently examined a number of low-order models which mimic finite extensibility (*102, 338*). The rigid dumbbell consists of two frictional beads held at a constant separation distance. In the string-bead dumbbell two beads are joined by a flexible string which transmits no force unless fully extended and serves only to place a limitation on the maximum separation of the beads. The finitely extendible non-linear elastic (FENE) dumbbell consists of two beads connected by a spring which is linear for small extensions but becomes non-linear at large extensions, having in fact an upper limit, $r = r_0$, on its extension (338):

$$F_{sp} = \frac{Kr}{1 - \left(\dfrac{r}{r_0}\right)^2} \qquad (0 < r < r_0). \tag{8.14}$$

The bead separation vector is r, and $r = |r|$.

The equilibrium distribution of bead separation distances in the FENE dumbbell is:

$$\Psi_e = \left[1 - \left(\frac{r}{r_0}\right)^2\right]^{b/2} \qquad (0 < r < r_0) \tag{8.15}$$

in which $b = Kr_0^2/kT$. The limit $K \to 0$ at fixed r_0 corresponds to the string-bead dumbbell, while the limit $r_0 \to \infty$ at fixed K yields the elastic dumbbell. The force-extension curve for FENE dumbbells is in fact rather similar to the inverse Langevin function which describes the behavior of freely jointed chains of finite contour length. The flow-induced stresses contributed by dilute suspensions of such species in a viscous medium have been calculated by Warner (338). The viscosity is constant at low shear rates, then begins to decrease in a characteristic region of shear rates, eventually approaching the power-law behavior

$$\frac{\eta - \eta_s}{\eta_0 - \eta_s} \propto |\dot{\gamma}|^{-1/3}. \tag{8.16}$$

Internal viscosity (Section 4) provides another possible source of shear-rate dependence. For sufficiently rapid disturbances, a spring-bead model with internal viscosity acts like a rigid body; for sufficiently slow disturbances it is flexible and indefinitely extensible. The analytical difficulties for coupled, non-linear spring-bead systems are equally severe in linear spring-bead systems with internal viscosity. Even the elastic dumbbell with internal viscosity has only been solved exactly in the limit of small ε (339), where ε is the ratio of internal friction coefficient to molecular (external) friction coefficient $\zeta_0 n$. For this case, the viscosity decreases with shear rate.

All these potential mechanisms for shear rate dependence—changes in excluded volume, changes in hydrodynamic interaction, finite extensibility and internal viscosity—can be made to predict viscosity-shear rate behavior which imitates more or less faithfully the forms observed experimentally. The choice among them must therefore be based on simultaneous comparisons with other equilibrium and viscoelastic properties.

It seems possible to rule out finite extensibility as an explanation of shear rate dependence in the viscosity, based simply on the equilibrium properties of polymer chains and the parallel between η and η' in their departures from η_0. Experimentally, the mean square end-to-end vector obeys Gaussian statistics in θ-solvents: $\langle r^2 \rangle / M = $ constant. If the spring constant K in FENE dumbbells is chosen to match this condition, then $K = 3kT/\langle r^2 \rangle$. The parameter b is therefore given by

$$b = \frac{3r_0^2}{\langle r^2 \rangle} \qquad (8.17)$$

which must be a large number, of the order of the number of main chain atoms in the molecule, in order for Gaussian statistics to be obeyed. The steady shear viscosity departs from $\eta_0 - \eta_s$ at a dimensionless shear rate β_0. The dynamic viscosity departs from $\eta_0 - \eta_s$ at a dimensionless frequency β_0'. The ratio of these values, obtained from the coefficient of the quadratic terms in power series expressions for viscosity in FENE dumbbells (338), is

$$\frac{\dot{\gamma}_0}{\omega_0} = \left[\frac{(2b+11)(b+5)}{2(4b+17)} \right]^{1/2} \qquad (8.18)$$

which, for $b \gg 1$, becomes

$$\frac{\dot{\gamma}_0}{\omega_0} = \frac{b^{1/2}}{2}. \qquad (8.19)$$

Accordingly, given the necessity from equilibrium coil dimensions that $b \gg 1$, the shear rate and frequency departures predicted by FENE dumbbells are displaced from each other. Moreover, the displacement increases with chain length. This is a clearly inconsistent with experimental behavior at all levels of concentration, including infinite dilution. Thus, finite extensibility must fail as a general model for the onset of nonlinear viscoelastic behavior in flexible polymer systems. It could, of course, become important in some situations, such as in elongational and shear flows at very high rates of deformation.

Similar arguments can be raised against internal viscosity as a primary cause of shear rate dependence in the viscosity. A recent review (339a) has shown that current many-bead theories predict differences in $\dot{\gamma}_0$ and ω_0 when ε is small and chosen to fit the oscillatory data. The approximate nature of these analyses leaves serious doubts about their predictions in steady-state flow, however. A new theory has been developed (339b) which addresses some of these problems, but its predictions about oscillatory and steady-shear behavior appear to be very similar to those of the earlier theories.

Both the many-bead and dumbbell models with internal friction predict limiting viscosities at high frequencies, η'_∞, and high shear rates η_∞. The theories predict that η'_∞ and η_∞ are related, such that

$$\frac{\eta_\infty - \eta_s}{\eta_0 - \eta_s} = 1 - \frac{\eta'_\infty - \eta_s}{\eta_0 - \eta_s} \qquad \text{[many bead (339b)]}$$

$$\frac{\eta_\infty - \eta_s}{\eta_0 - \eta_s} = 1 - \frac{5}{2} \frac{\eta'_\infty - \eta_s}{\eta_0 - \eta_s} \qquad \text{[dumbbell (339)]}.$$

Experimentally, $(\eta'_\infty - \eta_s)/(\eta_0 - \eta_s) \ll 1$ (93, 117), meaning that $(\eta_\infty - \eta_s)/(\eta_0 - \eta_s)$ should be only slightly less than unity, and shear rate dependence should practically disappear if only internal viscosity is considered. Such behavior is not of course observed.

A final piece of evidence against both finite extensibility and internal viscosity is provided by flow birefringence studies. One would expect each to produce variations in the stress optical coefficient with shear rate, beginning near the onset of shear rate dependence in the viscosity (307). Experimentally, the stress-optical coefficient remains constant well beyond the onset of shear rate dependence in η for all ranges of polymer concentration (18, 340).

It seems unlikely that these conclusions would be changed if the calculations were extended rigorously to models containing many beads and springs. The long (linear viscoeleastic) relaxation times of the undisturbed system clearly govern the onset of non-Newtonian behavior. If the distribution of configurations at equilibrium is Gaussian, then very few of the molecules will be greatly extended, and their finite lengths can have very little bearing on the relaxation time distribution. It is therefore difficult to see how finite extensibility could play a role at the early stages of non-linear behavior. Likewise, internal viscosity is essentially a local phenomenon, and seems quite unlikely to make its presence felt in the long relaxation processes. Thurston and Peterlin's approximation calculation for multiple bead systems with internal viscosity (118) contains this result, and the linear viscoelastic data of Massa et al. (93), for polystyrene in Arochlor agree very well with the resulting equations. It would therefore be very surprising to find that internal viscosity, which has practically no influence on the long relaxation times, could provide a mechanism for shear rate dependence in the viscosity, when experiments show that the onset of shear rate dependence is controlled by those same long relaxation times.

Changes in excluded volume and in intramolecular hydrodynamic interaction appear at the present time to be the only acceptable explanations for the onset of shear rate dependence in systems without appreciable intermolecular interactions. It seems likely that both internal viscosity and finite extensibility would assume importance only at much higher shear rates.

8.2.2.2. Intermolecular Contributions. Increasing concentration reduces the effects of excluded volume and intramolecular hydrodynamic on viscoelastic properties (Section 5). Internal viscosity and finite extensibilty have already been eliminated as primary causes of shear rate dependence in the viscosity. Thus, none of the intramolecular mechanisms, even abetted by an increased effective viscosity in the molecular environment, can account for the increase in shear rate dependence with concentration, *e.g.*, the dependence of power-law exponent on coil overlap $c[\eta]$ (Fig. 8.9). Changes in intermolecular interaction with increased shear rate seems to be the only reasonable source of enhanced shear rate dependence, at least with respect to the early deviations from Newtonian behavior and through a substantial portion of the power law regime.

In Williams' model the coil distortion produced by flow alters the intermolecular interaction and gives rise a shear rate dependence in the viscosity (*217*). According to this analysis, the potential which governs the total force between overlapping molecules decreases as the molecules are drawn out in the flow direction. Williams calculates $\eta(\dot{\gamma})$ using the flow-dependent segment density distribution for elastic dumbbells. A monotonically decreasing function of $\dot{\gamma}$ is obtained which is algebraically complicated but roughly obeys the form:

$$\frac{\eta}{\eta_0} = \left[\frac{1}{1 + (\tau_0 \dot{\gamma})^2}\right]^{1/2}. \tag{8.20}$$

The time constant τ_0 is the relaxation time of the elastic dumbbell, which would be $(\eta_0 - \eta_s) M/cRT$ if the dumbbells were independent. Williams makes the dimensional argument described earlier (Part 6) in favor of a different form, the result being

$$\tau_0 = F(c)\frac{\eta_0 - \eta_s}{c^2 RT} \tag{8.21}$$

in which $F(c)$ depends only upon the equilibrium thermodynamic properties of the solution. It is assumed to be independent of polymer molecular weight and at most only a weak function of concentration.

This treatment assumes that the forces between molecules in relative motion are related directly to the thermodynamic properties of the solution. The excluded volume does indeed exert an indirect effect on transport properties in dilute solutions through its influence on chain dimensions. Also, there is probably a close relationship between such thermodynamic properties as isothermal compressibility and the free volume parameters which control segmental friction. However, there is no evidence to support a direct connection between solution thermodynamics and the frictional forces associated with large scale molecular structure at any level of polymer concentration.

The theory associates the onset of shear rate dependence with coil distortion, and is thus consistent with the close relationship which is observed between η vs $\dot\gamma$ and η' vs ω. It does not predict any dependence of the $f(\beta)$ form on concentration, contrary to observations in the low to moderate concentration range. Also, in this same range the characteristic relaxation time obeys the form suggested by the Rouse theory, rather than $(\eta_0 - \eta_2)/c^2 R T$. The Williams theory, in fact, describes highly concentrated solutions and melts somewhat better than the moderately concentrated but unentangled solutions for which it was developed. This may mean that the formalism provides a reasonable attack on flow properties at high concentrations, but without the need to invoke chain entanglement directly.

Further examination of the Williams approach seems called for, both to improve the method for estimating parameters such as the relaxation time, and to clarify the relationship between the intermolecular potential form and non-thermodynamic frictional forces. The method might provide a fairly unified description of non-linear flow porperties if a suitable "potential function" for large scale molecular friction were found. Aside from the Williams work, there have been no theoretical studies dealing with η vs. $\dot\gamma$ at low to moderate concentrations. The systematic changes in the master curve $f(\beta)$ with coil overlap $c[\eta]$ are thus without explanation at the present time.

Experimental viscosity-shear rate curves at high concentrations turns out to be rather similar to an expression for non-Newtonian viscosity derived from the Eyring's activated complex theory for the transport properties of liquids (*341*):

$$\frac{\eta}{\eta_0} = \frac{\sinh^{-1}\dot\gamma\,\tau_0}{\dot\gamma\,\tau_0}. \tag{8.22}$$

The derivation of this equation makes it seem more appropriate for small molecule liquids or suspensions of hard spheres than for interpenetrating random coil molecules however. Indeed, a somewhat modified version of the equation has approximately the form observed for $\eta(\dot\gamma)$ in concentrated suspensions of monodisperse spheres. In this case $1/\tau_0$ turns out to be of the order of the rotational diffusion of the spheres, $\dfrac{\eta_0\,a^3}{kT}$ (*342, 343*), the parameter a being the radius of of the spheres.

The Eyring analysis does not explicity take chain structures into account, so its molecular picture is not obviously applicable to polymer systems. It also does not appear to predict normal stress differences in shear flow. Consequently, the mechanism of shear-rate dependence and the physical interpretation of the characteristic time τ_0 are unclear, as are their relationships to molecular structure and to cooperative configurational relaxation as reflected by the linear viscoelastic behavior. At the present time it is uncertain whether the agreement with experiment is simply fortuitous, or whether it signifies some kind of underlying unity in the shear rate dependence of concentrated systems of identical particles, regardless of their structure and the mechanism of interaction.

A master curve appropriate for concentrated systems has been derived from the assumption that shear rate dependence is caused by a progressive decrease in the steady-state entanglement density (227, 344). Linear viscoelastic behavior indicates that each molecule at equilibrium participates in a number of potential coupling arrangements with other molecules lying within its pervaded volume. Random thermal motion continually changes the configuration of each chain, new potential couples being formed while old ones are lost. If the average number of couples per chain is large, the fluctuations in coupling density will be small and the number of couples per chain will be roughly constant in time. This is the equilibrium situation envisioned in most relaxing network theories.

Macroscopic deformation imposes a systematic relative motion on the molecules. If each is entangled with a large number of other molecules, its center of gravity moves with the macroscopic velocity at its location since the effects of entanglement drag on its trajectory cancel. If the motion is steadily continued, each molecule sees a constant stream of potential coupling partners passing through its pervaded volume. If the flow is slow enough, random thermal motion can establish new coupling arrangements with incoming partners as previous partners leave. Thus, for sufficiently slow flows (or rapid but sufficiently small deformations), the topological state of the system remains essentially the same as that at equilibrium.

For higher deformation rates, the time that potential partners remain within the pervaded volume of a molecule will decrease. The time of passage τ_p is an external time, governed by the macroscopic deformation rate and the trajectory of each potential coupling partner. The time to form an entanglement τ_e, on the other hand, is to be viewed as an internal time, governed by the rate of large-scale configurational relaxation and of the order of the configurational memory time (Section 4) for the chains. The coupling probability is assumed to depend on the relative values of τ_p and τ_e, approaching the equilibrium probability when $\tau_e \ll \tau_p$ and vanishing when $\tau_e \gg \tau_p$. Accordingly, the steady-state entanglement density of the system is expected to be a decreasing function of the deformation rate.

A simplified model of this situation has been used to calculate viscosity-shear rate behavior. The steady-state power dissipation per unit volume $\dot{E}_v$ was calculated as the sum of the power dissipations by individual molecules, each molecule being treated as if it interacted independently with the surrounding medium. Entanglement probability was represented as a step function. Interactions with elements in the medium which pass through the pervaded volume of the molecule (taken to be a sphere of unspecified R_s) in a time less than τ_e were omitted, while those for which the passage time was greater than τ_e were given their full weight. Effects due to coil rotation and to deformation of the pervaded volume by the frictional forces were ignored. [According to a recent discussion (212), entanglement drag tends to align the sections of chain between coupling points, but produces considerably less overall distortion of the coil than drag in a continuous medium of the same macroscopic viscosity.]

It was also assumed that internal times such as τ_e are proportional to the macroscopic viscosity: $\tau_e = \eta/\eta_0 (\tau_e)_0$; $(\tau_e)_0$ or τ_0 is the mean entanglement time at equilibrium. The viscosity was then calculated from the expression $\eta = \dot{E}_v/\dot{\gamma}^2$, which is an identity for simple shear flow at steady state. The result is an equation for the viscosity master curve in monodisperse systems (227):

$$\frac{\eta}{\eta_0} = g^{3/2} h \tag{8.23}$$

in which

$$g(\theta) = \frac{2}{\pi}\left[\cot^{-1}\theta + \frac{\theta}{1+\theta^2}\right], \tag{8.24}$$

$$h(\theta) = \frac{2}{\pi}\left[\cot^{-1}\theta + \frac{\theta(1-\theta^2)}{(1+\theta^2)^2}\right], \tag{8.25}$$

and

$$\theta = \frac{\eta}{\eta_0}\frac{\dot\gamma\tau_0}{2}. \tag{8.26}$$

The functions g and h represent two separate effects of disentanglement. First, each molecule has fewer couples with the medium than at equilibrium, couples being lost preferentially at the coil periphery in the direction of the velocity gradient. The power dissipation is therefore less by the factor $1-h$ than would be the case if all couples available at equilibrium were acting. Second, the medium itself has fewer couples than at equilibrium. Those couples remaining between the molecule and the medium exert drag forces which are smaller by the factor $g^{3/2}$ than would be the case if all couples present at equilibrium were acting. The function $1-g$ is the fractional reduction in the entanglement density ($g = E/E_0$) under the existing condition of flow. Equations (8.23)–(8.26) define an implicit expression for the master curve η/η_0 vs $\dot\gamma\tau_0$, the reduced viscosity η/η_0 being also present in the arguments of the functions g and h. Appropriate calculations yield a result which turns out to be practically indistinguishable from Eyring's equation (341).

Bueche (347) has analyzed a similar disentanglement model, evaluating in somewhat greater detail the coupling probabilities and taking into account the expected differences in τ_e for entanglements of different complexities. Again the form is very similar to Eqs. (8.22) and (8.23).

Table 8.1 compares the asymptotic behavior at high rates of the various master curves for monodisperse systems. Their limiting forms

Table 8.1. Behavior of theoretical master curves for viscosity at high shear rates

Investigator	Asymptotic form of η/η_0
Williams (217)	$1/\dot\gamma\tau_0$
Eyring (341)	$\ln\dot\gamma\tau_0/\dot\gamma\tau_0$
Graessley (227)	$1/(\dot\gamma\tau_0)^{9/11}$
Bueche (347)	$1/(\dot\gamma\tau_0)^{6/7}$

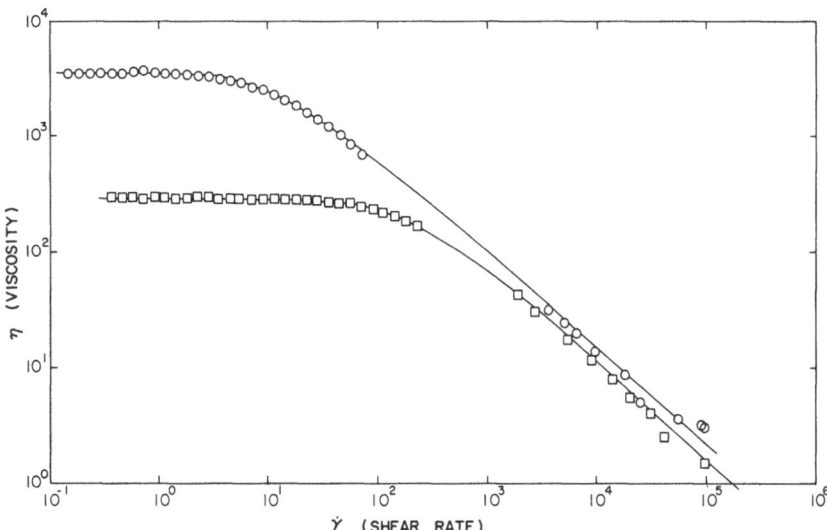

Fig. 8.15. Viscosity vs shear rate in concentrated solutions of narrow distribution polystyrene. The solvent in n-butyl benzene, the concentration is 0.300 gm/ml and the temperature is 30° C. The symbols are ○ for $M = 860000$ and □ for $M = 411000$ at low shear rates (155) and at high shear rates (346). The solid line for $M = 860000$ is the master curve for monodisperse systems from Graessley (227). The solid line for $M = 411000$ is the master curve from Ree-Eyring (341). Either master curve fits data for both molecular weights

are reached after a decrease of about one decade in η/η_0. It is difficult to distinguish among them experimentally, partly because data becomes less certain at very high shear rates and partly because residual polydispersity would probably obscure the relatively small differences anyway. In any case, the shape of the curves between the Newtonian and high shear rate regions, where many data are available, is roughly the same in the four equations and in good agreement with experiment (155, 346) (Fig. 8.15).

Theories based on the uniformly effective medium have the practical advantage that they can be extended quite easily to polydisperse systems (227). Viscosity master curves can be predicted from the molecular weight distribution, for example. The only new assumption is that the entanglement time at equilibrium for a chain of molecular weight M in a polydisperse system has the form suggested by the Rouse theory (15):

$$\tau_r(M) = \frac{6}{\pi^2} \frac{\eta_0 \bar{M}_w}{cRT} \left(\frac{M}{\bar{M}_w}\right)^2 = \tau_R \left(\frac{M}{M_w}\right)^2. \tag{8.27}$$

That is, at zero shear rate

$$\tau_e(M) = \tau_0 \left(\frac{M}{M_w}\right)^2 \tag{8.28}$$

in which τ_0 is now a communal time constant, evaluated [as suggested by Eq. (8.27)] for a monodisperse system with the same molecular weight as $\bar{M}_w$ of the polydisperse system.

The agreement with experimental master curves in polydisperse systems turns out to be fairly good (345), although some significant differences may exist for very broad molecular weight distributions (348). Values of τ_0 derived from systems of different polydispersities also correlate well in terms of $\bar{M}_w$ (345). One hesitates to drawn firm conclusions about the validity of the underlying concept of chain disentanglement from this agreement, however. For example, Middleman (349) evaluates the master curve for a polydisperse system simply as a weighted sum of monodisperse master curves. He uses the Rouse form for the weighting factor [Eq. (8.27)] but introduces no explicit considerations of mechanism. The predicted shapes are not in quite as good agreement with experiment as those based on the disentanglement hypothesis, and the values of τ_0 obtained correlate less well in terms of $\bar{M}_w$ (345). However, the general features of the predicted curves are very similar. What does seem to be clear is that entanglement is an acceptable explanation of shear rate dependence and that the Rouse weighting factor for individual contributions, $(\bar{M}/\bar{M}_w)^2$, is roughly correct in polydisperse systems.

Unfortunately, the energy dissipation method upon which these calculations are based is only applicable to the evaluation of viscosity in steady deformations. The method does not lend itself to an extension of the disentanglement model to other components of the stress or to other types of deformation history.

8.3. First Normal Stress Function

The first normal stress function Ψ_1 is defined as

$$\Psi_1(\dot{\gamma}) = \frac{p_{11} - p_{22}}{\dot{\gamma}^2} = \frac{N_1}{\dot{\gamma}^2}. \tag{8.29}$$

The properties of $\Psi_1(0)$ for narrow distribution polymers have already been discussed in Section 5. The behavior of $\Psi_1(\dot{\gamma})$ at higher shear rates has only been determined for a few systems of well-characterized molecular structure. The experimental problems are more difficult than in the case of $\eta(\dot{\gamma})$, so the conclusions here must be regarded as somewhat more tentative. Experimentally, $\Psi_1(\dot{\gamma})$ and $\eta(\dot{\gamma})$ depart from their zero shear values within the same range of shear rates (172). Shear rate sensitivity is much smaller when N_1 is expressed as a function of shear stress (350).

$$N_1 = 2 J(\dot{\gamma}) \sigma^2. \tag{8.30}$$

At the limit of low shear rates $J(\dot{\gamma})$ approaches J_e^0 according to the Coleman-Markovitz relation [Eq. (3.21)] and the experimental results in Section 5. In many systems $J(\dot{\gamma})$ is a decreasing function of shear rate (351). However, Graessley and Segal (352) found that $J(\dot{\gamma})$ increased with $\dot{\gamma}$ in concentrated solutions of polystyrene with narrow molecular weight distribution, while decreasing with $\dot{\gamma}$ in broad distribution samples. Nagasawa and co-workers (198, 199) have found similar behavior in poly(α-methyl styrene) systems, and Uy and Graessley (195) in polyvinyl acetate, although in the latter study the shear rate dependence of $J(\dot{\gamma})$ for narrow distributions was quite small in all cases.

Such behavior is qualitatively understandable in terms of partial disentanglement in steady shear flow. In highly entangled systems ($cM > \varrho M_c'$) J_e^0 is of the form (Section 5):

$$J_e^0 \propto \frac{M}{cE_0RT} \tag{8.31}$$

in which E_0 is $M/(M_e)_{\text{soln}}$, the number of entanglement points per molecule at equilibrium. According to the entanglement theory of $\eta(\dot{\gamma})$, the steady state number of entanglements at shear rate $\dot{\gamma}$ is reduced by the factor g [Eq. (8.24)], so that

$$J(\dot{\gamma}) \propto \frac{M}{cERT} = \frac{M}{cgE_0RT} \tag{8.32}$$

or

$$J(\dot{\gamma}) = J_e^0/g(\dot{\gamma}\tau_0). \tag{8.33}$$

Since g is a monotonically decreasing function of shear rate, departing from unity at the onset of shear rate dependence on η, $J(\dot{\gamma})$ should show positive curvature (a montonic increase with shear rate) for systems of narrow molecular weight distribution. The shear rate dependence shown by Nagasawa's data (Fig. 8.16) is qualitatively similar but actually somewhat greater than that predicted with values of g from Eq. (8.24).

The effect of polydispersity on $J(\dot{\gamma})$ is also consistent with a disentanglement mechanism. The largest molecules in the system contribute the longest relaxation times to the linear viscoelastic spectrum, and entanglements are lost from large molecules at lower shear rates. The long relaxation time processes are therefore removed, progressively reducing the breadth of the relaxation time distribution with increasing shear rate and tending thereby to reduce the apparent compliance function $J(\dot{\gamma})$. If shear-induced narrowing of the relaxation time distribution dominates in polydisperse systems, the function $J(\dot{\gamma})$ will decrease with shear rate, as observed. Tanaka, Yamamoto and Takano (353) have developed a theory for predicting non-linear behavior from flow-induced changes in the relaxation spectrum. They concluded that the behavior of $J(\dot{\gamma})$ depends on the shape of the terminal linear viscoelastic spectrum, decreasing with shear rate when the spectrum is broad and increasing when it is narrow.

Such qualitative agreement should not be interpreted too broadly in favor of disentanglement however. Sakai et al. (199) have also observed that $J(\dot{\gamma})$ increases with $\dot{\gamma}$ even in narrow distribution systems of lower concentration, that is,

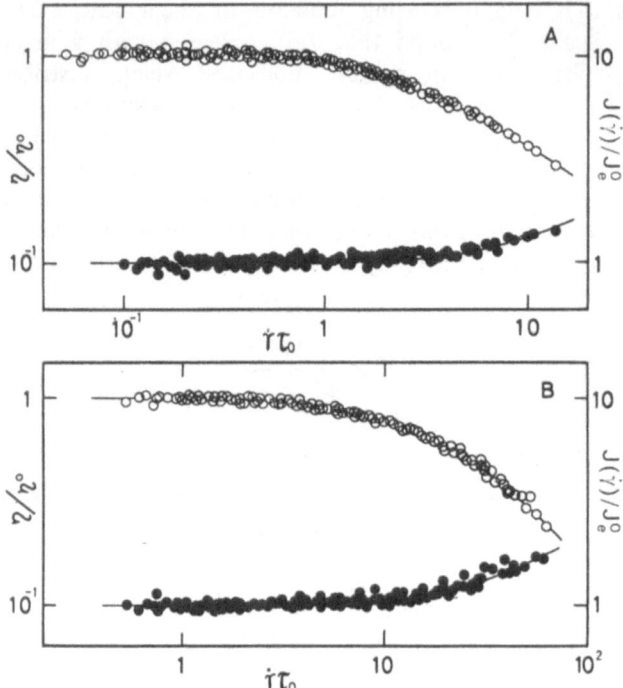

Fig. 8.16. Superimposed plots of $\eta(\dot{\gamma})/\eta_0$ and $J(\dot{\gamma})/J_e^0$ as functions of shear rate for poly(α-methyl styrene) solutions (199). Curve A is for data at low concentrations and molecular weights; curve B is for data at high concentrations and molecular weights. Data have been shifted along the shear rate axis to produce superposition. [Reproduced from Macromolecules **5**, 791 (1972).]

where J_e^0 still conforms to the Rouse expression; in this case the argument leading to Eq. (8.33) is no longer valid. Also, certain dilute solution models, such as FENE dumbbells, predict an increase in $J(\dot{\gamma})$ with shear rate,

$$\text{Rouse model:} \qquad J(\dot{\gamma}) = \text{constant} \quad (\text{all } \dot{\gamma}), \qquad (8.34)$$

$$\text{Entanglement model:} \qquad J(\dot{\gamma}) \propto \dot{\gamma}^{2/11} \qquad (\dot{\gamma}\tau_0 \gg 1), \qquad (8.35)$$

$$\text{FENE dumbbell model (338):} \quad J(\dot{\gamma}) \propto \dot{\gamma}^{2/3} \qquad (\dot{\gamma}\tau_0 \gg 1). \qquad (8.36)$$

Although the FENE model has been shown earlier to be inappropriate for polymer systems on other grounds, it nevertheless illustrates an instance in which $J(\dot{\gamma})$ increases with shear rate naturally and without the need for postulating structural changes in the system.

8.4. Second Normal Stress Function

Values of $p_{22} - p_{33} = N_2$ appear to be negative and approximately 10–30% of N_1 in magnitude (82). The conventional bead-spring models yield $N_2 \equiv 0$. Indeed, N_2 in steady shear flow is identically zero for all free draining models, regardless of the force-distance law in the connectors (102a). Thus, finite extensibility and, by inference at least, internal viscosity do not in themselves provide non-zero N_2 values. Bird and Warner (354) have recently analyzed the rigid dumbbell model with intramolecular hydrodynamic interaction, the latter represented by the Oseen approximation. In this case N_2 turns out to be non-zero but positive.

The bead-connector models have only been used in calculations for systems at infinite dilution, and N_2 has been measured in moderately concentrated solutions. It is therefore possible that N_2 is associated only with intermolecular effects. The Lodge and Yamamoto network models (Section 6) introduce a kind of intermolecular interaction, but they too yield $N_2 \equiv 0$. The Williams model (219) includes intermolecular effects in another manner. In this case N_2 is negative (although changing sign at high shear rates), but much larger than N_1, the latter approaching zero as $\dot{\gamma}^4$, contrary to all observations. Thus, there is no compelling reason, molecular or continuum, to expect N_2 to have any special magnitude or sign with respect to N_1.

It is interesting to examine the bead-spring models to see what flow-induced configurational changes would be required in order to develop N_2 values of the proper magnitude and sign. In the Rouse model, the components of the stress tensor are related directly to averages of the internal coordinates of the beads. For the simplest case of the elastic dumbbell:

$$p_{ij} = -P_0 \delta_{ij} + \frac{12\nu kT}{\langle r^2 \rangle}[\langle x_i x_j \rangle - \langle x_i x_j \rangle_0] \tag{8.37}$$

in which x_1, x_2, and x_3 are the bead components relative to the center of gravity of the dumbbell, ν is the number of dumbbells per unit volume, $\langle r^2 \rangle$ is the equilibrium mean square bead separation distance, and $\langle x_i x_j \rangle$ and $\langle x_i x_j \rangle_0$ are averages evaluated during flow and at equilibrium respectively.

$$\langle x_i x_j \rangle = \int\!\!\!\int\!\!\!\int_{-\infty}^{\infty} x_i x_j \, \Psi \, dx_1 dx_2 dx_3 , \tag{8.38}$$

$$\langle x_i x_j \rangle_0 = \frac{\langle r^2 \rangle}{12} \quad (i=j), \tag{8.39}$$

$$= 0 \qquad (i \neq j).$$

The normal stress functions are

$$N_1 = \frac{12 \, vkT}{\langle r^2 \rangle} [\langle x_1^2 \rangle - \langle x_2^2 \rangle], \tag{8.40}$$

$$N_2 = \frac{12 \, vkT}{\langle r^2 \rangle} [\langle x_2^2 \rangle - \langle x_3^2 \rangle]. \tag{8.41}$$

The positive value of N_1 thus reflects an expansion of the distribution of configurations in the flow direction, relative to the distribution in the direction of the velocity gradient. Negative N_2 values imply an *effective* contraction of the distribution along the direction of the velocity gradient relative to the 3 direction.

In the Rouse model the distribution of configurations in the 2 and 3 directions are independent of flow in the 1 direction, so N_2 is identically zero. Thus, one might speculate than N_2 would be negative if the frictional coefficient of a bead depended upon x_2 such that ζ decreased as $|x_2|$ increased. Such an effect could result from a systematic depletion of entanglements from segments with large values of $|x_2|$. In Graessley's theory (227), the ratio h/g is literally the effective reduction in $\langle x_2^2 \rangle$ for the frictional resistances of the coil due to a systematic loss of entanglements at its periphery. Thus, with Eq.(8.41), one might write

$$N_2 = - vkT(1 - h/g) \tag{8.42}$$

and, for $\dot{\gamma} \tau_0 \ll 1$, using Eqs.(8.24) and (8.25),

$$N_2 = - vkT \left[\frac{4}{\pi} \left| \frac{\dot{\gamma} \tau_0}{2} \right|^3 \right] \propto - |\dot{\gamma}|^3. \tag{8.43}$$

The sign of N_2 is correct but the dependence on $|\dot{\gamma}|^3$ at low shear rates is rather contrary to expectations.

Wales and Philippoff (345a) have suggested recently that negative values of N_2 may be caused by differences in segmental diffusion coefficients parallel and transverse to the chain direction. The beads in the spring-bead models are conventionally treated as spheres, so no directional bias is imposed. If the chain segments are non-spherical, and the parallel and transverse diffusion coefficients of the beads, D_p and D_t, are calculated from the average spatial orientation of the segments, finite values of N_2 are obtained:

$$\frac{N_2}{N_1} = \frac{D_t - D_p}{5 D_t + D_p}. \tag{8.44}$$

When the segments behave as long cylinders, $D_p = 2D_t$, and:

$$\frac{N_2}{N_1} = -1/8. \tag{8.45}$$

Even considering this rather interesting suggestion of Wales and Philippoff (*354a*), the molecular origins of N_2 remain uncertain at the present time. One might hope that experimental data on this property in well-characterized systems, particularly on its behavior at low shear rates, will in the future provide some guidance.

8.5. Non-linear Response in Time-dependent Simple Shear Flows

Studies have been made of the stresses produced in several non-steady flow histories. These include the buildup to steady state of σ and $p_{11} - p_{22}$ at the onset of steady shearing flow (*355–358*); relaxation of stresses from their steady state values when the flow is suddenly stopped (*356–360*); stress relaxation after suddenly imposed large deformations (*361*); recoil behavior when the shear stress is suddenly removed after a steady state in the non-linear region has been reached (*362*); and parallel or transverse oscillations superimposed on steady shearing flow (*363–367*). Experimental problems caused by the inertia and compliance of the experimental apparatus are much more severe than in steady state measurements (*368, 369*). Quantitative interpretations must therefore still be somewhat tentative. Nevertheless, the pattern of behavior emerging is suggestive with respect to possible molecular flow mechanisms.

8.5.1. Stress Development at the Onset of Steady Shearing Flow (*355–358*)

At sufficiently low shear rates the viscosity is constant. If a constant shear rate in this range is imposed, the shear stress should grow monotonically to its steady state value $\sigma(\infty)$ according to the equation from linear viscoelasticity:

$$\sigma(t) = \sigma(\infty) \, \frac{\int\limits_{-\infty}^{\infty} \tau H(\tau) \left[1 - e^{-t/\tau}\right] d\ln\tau}{\int\limits_{-\infty}^{\infty} \tau H(\tau) \, d\ln\tau} \tag{8.46}$$

where

$$\dot{\gamma}(t) = 0 \quad (t < 0)$$

$$\dot{\gamma}(t) = \dot{\gamma} \quad (t > 0)$$

$$\sigma(\infty) = \eta_0 \dot{\gamma}.$$

At higher shear rates the steady-state viscosity $\sigma(\infty)/\dot{\gamma}$ begins to depend on $\dot{\gamma}$. If a constant shear rate in this range is imposed, the shear stress rises beyond its steady state value, passes through a maximum before eventually approaching the steady state, $\sigma(\infty) = \eta(\dot{\gamma})\dot{\gamma}$. The stress at the maximum $\sigma(t_m)$ grows in relation to $\sigma(\infty)$ as $\dot{\gamma}$ increases. The time to reach the maximum t_m is found to be inversely proportional to shear rate:

$$t_m \dot{\gamma} = \text{constant}. \tag{8.47}$$

The total shear at the stress maximum γ_m is $\dot{\gamma} t_m$, so for a given polymer γ_m is independent of $\dot{\gamma}$, and has a value which is typically in the range of 2.0–3.0 shear units. The normal stress displays a similar overshoot behavior, although γ_m and the time to reach steady state appear to be greater for p_{11-22} than for σ.

8.5.2. Stress Decay at the Termination of Steady Shearing Flow (356–360)

At sufficiently low shear rates the shear stress should decay at the termination of steady state shear flow according to the equation from linear viscoelasticity:

$$\sigma(t) = \sigma(0) \ \frac{\displaystyle\int_{-\infty}^{\infty} \tau H(\tau) e^{-t/\tau} d \ln \tau}{\displaystyle\int_{-\infty}^{\infty} \tau H(\tau) d \ln \tau} \tag{8.48}$$

where

$$\dot{\gamma}(t) = \dot{\gamma} \quad (t < 0)$$

$$\dot{\gamma}(t) = 0 \quad (t > 0)$$

$$\sigma(0) = \eta_0 \dot{\gamma}.$$

Accordingly. plots of $\sigma(t)/\sigma(0)$ vs t from different shear rates should super-impose. Experimentally the curves do not superimpose when the shear rate is in the non-Newtonian region, the initial rate of relaxation being increasingly more rapid for higher shear rates. The normal stress decays more slowly than shear stress, but behaves similarly with respect to the effect of previous shearing flow in the non-Newtonian region.

8.5.3. Stress Relaxation from Sudden Strains of Large Amplitude (361)

For shear strains of sufficiently small amplitude, the response is linear and the shear stress is governed by the stress relaxation modulus:

$$\sigma(t) = \gamma_0 G(t). \tag{3.9}$$

For shear strains greater than approximately 2 the ratio $\sigma(t)/\gamma_0$ for a concentrated polystyrene solution was reduced at all observable times. For the large strains, relaxation proceeded more rapidly at short times, but at longer times the residual stress decayed with about the same time dependence as that in the linear viscoelastic region.

8.5.4. Shear Recovery after Steady Shearing Flow (362)

The total amount of shear recoil after steady state shear flow at sufficiently low shear rates is related to J_e^0:

$$\gamma_r = J_e^0 \sigma_0 \tag{3.17}$$

where

$$\sigma(t) = \sigma_0 \qquad (t < 0)$$

$$\sigma(t) = 0 \qquad (t > 0)$$

$$\sigma_0 = \eta_0 \dot{\gamma}.$$

When the shear rate at steady state is in the non-Newtonian region for the fluid, the recoil is smaller than $J_e^0 \sigma_0$ for both broad and narrow molecular weight distributions (362).

8.5.5. Small Oscillations Superimposed on Steady Shearing Flow (363–367)

The dynamic moduli for infinitesimal superimposed deformations parallel and transverse to the flow direction in steady shearing flow should be unaffected by flow if the shear rate is sufficiently small. According to the theory of simple fluids, the superimposed dynamic moduli for shearing flows in the non-Newtonian region must change in order to conform with the relations (370, 371 :

$$\lim_{\omega \to 0} \frac{G_\perp''(\dot{\gamma},\omega)}{\omega} = \eta(\dot{\gamma}), \tag{8.49}$$

$$\lim_{\omega \to 0} \frac{G_\parallel''(\dot{\gamma},\omega)}{\omega} = \eta(\dot{\gamma})\left[1 + \frac{d\log\eta}{d\log\dot{\gamma}}\right]. \tag{8.50}$$

The loss moduli at low frequencies are indeed reduced at high shear rates (363–367), as implied by these equations. No conclusions about the behavior of the storage moduli can be drawn from simple fluid theory. However, $G'(\dot{\gamma},\omega)$ and $G_\parallel'(\dot{\gamma},\omega)$ are also reduced at high shear rates, and $G_\parallel'(\gamma,\omega)$ may even become negative at low frequencies (365).

8.5.6. Discussion

The slower rise and decay of normal stress transients compared to shear stress arises quite simply and directly from the polydispersity of relaxation times (78), and probably has no direct bearing on entanglement mechanisms *per se.* Likewise, the depression of the superimposed moduli at low frequencies follows from rather non-specific continuum models, the loss moduli by Eqs. (8.49) and (8.50) from the simple fluid model, and the storage moduli from the following properties of the more specific but still quite general BKZ model (366, 372):

$$\lim_{\omega \to 0} \frac{2 G'_\perp (\dot\gamma, \omega)}{\omega^2} = \frac{N_1}{\dot\gamma^2}, \tag{8.51}$$

$$\lim_{\omega \to 0} \frac{2 G'_\parallel (\dot\gamma, \omega)}{\omega^2} = \frac{N_1}{\dot\gamma^2} \left[1 + \frac{d \log N_1}{d \log \dot\gamma} - 1 \right]. \tag{8.52}$$

According to Eq. (8.52), negative values of $G'_\parallel$ at low frequencies occur if N_1 becomes proportional to a power of shear rate less than one, a condition which should be attainable at moderate shear rates in concentrated solutions of polymers with narrow molecular weight distribution. Thus, any molecular theory which predicts η and N_1 as functions of shear rate, and which is also consistent with the BKZ and simple fluid theories should automatically yield superimposed moduli which satisfy Eqs. (8.49)–(8.52) without special assumptions about entanglement mechanisms.

Stress overshoot at the onset of steady shearing flow and accelerated relaxation at its termination are properties which are not automatic consequences of general continuum theories. They can be interpreted as resulting from a partial decoupling of the system which begins at a characteristic shear strain at the onset of rapid shearing flow, and a history-dependent re-establishment of the equilibrium degree of coupling after flow is stopped. In the integral continuum models this behavior corresponds to a flow history dependence in the memory function. The behavior appears to be inherently unsymmetrical in that *the initial decoupling depends explicitly on the extent of strain, while re-coupling depends only on time.* The initial relaxation of a sheared system is rapid because the system is partially decoupled and therefore presents a reduced resistance to configurational rearrangement. As relaxation proceeds, the structure returns to its equilibrium state, the coupling builds up and the resistance to further relaxation increases. One might expect the last traces of stress to fade at essentially the linear viscoelastic rate for the system, although evidence on this point is still unclear (373).

Lodge and Meissner have recently examined in detail the stresses during start-up of steady elongational and shear flows (374). The Lodge equation [Eq. (6.15)], with its flow-independent memory function, described the build-up of stress rather well (even for rapid deformations) until a critical strain was reached. Beyond the critical strain (which differed somewhat in shear and elongational deformations), a history-dependent memory function was required. Similar behavior has been noted in the rheologically more complicated property of die-swell. Bagley and co-workers have studied the effect of capillary length on the ratio of extrudate diameter to capillary diameter at constant flow rate (375). For short capillaries the extrudate diameter is large, but it decreases with increasing length and appears to approach a limiting value for long capillaries. They demonstrated that the criterion for reaching the asymptotic diameter at each shear rate was not the residence time in the capillary, but rather the total amount of deformation experienced during passage through the capillary.

Stratton and Butcher have studied the effect of interrupted flow on response (376). If a steady flow is stopped and then restarted quickly, the overshoot is missing. For longer periods of rest the overshoot gradually reappears, showing that during rest the equilibrium structure of the system is slowly re-established.

Great care is necessary in introducing a dependence of memory function relaxation times on the previous flow history. For example, if the relaxation times are expressed simply in terms of past strain rates alone, then accelerated relaxation at the termination of steady-state flow may indeed be accommodated satisfactorily. However, stress data at the onset of flow appear to be fundamentally inconsistent with this kind of modification (377). Also the ability of the modified constitutive equation to handle small but rapid oscillatory deformations [such as are used to measure $G'(\omega)$ and $G''(\omega)$] may be seriously compromised. Strain rate magnitudes may grow very large for high frequencies even though strain amplitudes remain small. Information on the properties and limitations of integral constitutive equations is accumulating rapidly (378, 379), but a detailed summary of current work is beyond the scope of the present review.

Ziabicki (380) has pointed out that flow-history dependence of the stress translates in molecular terms to a flow-history dependence of the distribution of configurations, with current stress depending directly only on the current configuration. Strain criteria for evoking a history dependence in the memory function imply that deformation in itself can alter or even obliterate the contribution of past states to the current configuration. [Among current constitutive equations of fairly specific form, the network-rupture model of Tanner (381) corresponds most closely to this idea.] Flow-induced changes in entanglement coupling or topological state serve as a useful concept for discussing and rationalizing many aspects of non-linear behavior (382). The incorporation of such ideas into a quantitative constitutive theory is of course an eagerly sought goal.

9. Conclusions

The conclusions of this review can be separated into those which concern the experimental rheology of networks and concentrated systems of narrow distribution polymers and those related to the ability of molecular theories to rationalize and predict these observations.

Rheological behavior at low to moderate concentrations parallels that for particulate suspensions in its dependence on the Simha interaction parameter $c[\eta]$. The relative viscosity η_0/η_s is primarily a function of $c[\eta]$. The steady state compliance J_e^0 lies in the range predicted by the conventional spring-bead models, with residual variations depending on $c[\eta]$. The shear rate sensitivity of viscosity, judged by the power law exponent, increases rapidly through this region and depends principally on $c[\eta]$. There seems to be little necessity to postulate specific interactions such as chain entanglement to rationalize most behavior in this regime.

In concentrated solutions and melts the behavior is controlled by Bueche's interaction parameter cM and the magnitude of the local friction factor ζ_0. The appearance of a plateau in the relaxation modulus, the onset of $M^{3.4}$ dependence in η_0, and the change in form of J_e^0 from M/cRT to $1/c^2RT$ all depend on cM. The characteristic molecular weights associated with G_N^0, η_0 and J_e^0 in the un-diluted state — M_e, M_c, and M_c' respectively — are related to each other in approximately the same way for a number of different polymers. The characteristic molecular weights are all inversely proportional to concentration and virtually independent of temperature and the choice of solvent. With some exceptions the values of M_e, associated with the average molecular weight between entanglements, correspond to roughly the same contour length for different polymers. For those relatively few cases in which the chemical structure of crosslinked networks has been characterized in detail, the values of M_e obtained from the equilibrium modulus are similar to values obtained from the plateau modulus for the same polymer without crosslinks. All these properties, which point clearly to a universal physical interaction between chain molecules, are documented rather completely in non-polar, non-crystallizable polymers. Information is much less complete for systems of highly polar, stiff, and crystallizable chains.

The factors affecting the transition between low and high concentration behavior, taking place roughly in the range 10–20% polymer but depending strongly on chain length and the nature of both polymer and solvent, have been studied very little and are not well understood.

A number of suggestions about the nature of the entanglement interaction have been put forward. *The only acceptable interpretation appears to be that it is a direct consequence of the uncrossability of chain contours.* Uncrossability imposes topological restrictions which reduce the number of paths available for configurational relaxation in a system of real chains, compared to a hypothetical system of crossable (ghost) chains. The long relaxation times, those governing large scale configurational change, are therefore larger in the system of real chains. The same effect reduces the number of available configurations in deformed networks of real chains. The result of such permanent topological classification has been shown invariably to increase the equilibrium modulus, making it larger than that of an equivalent network of ghost chains.

The large scale rope-like interactions pictured by Bueche are crude representations, but certainly more realistic than interactions which depend on local kinks, rings, or other special sources of contact friction or molecular roughness. We are however a long way from being able to associate entanglement effects with specific attributes of the topological arrangement. It would be premature, for example, to associate on a one-to-one basis each "equivalent crosslink" of the topological modulus contribution with one of Bueche's interlacing loops.

Unfortunately, theories on the shift of relaxation times caused by entanglement—essentially explanations of the $M^{3.4}$ dependence of η_0—are still quite arbitrary and inelegant. The early qualitative ideas of Treloar appear to be basically correct, but quantitative theories have not progressed far beyond Bueche's original analysis of 20 years ago. Not surprisingly, the theory of the terminal spectrum in highly entangled systems is also less than satisfactory. Very little progress has been made beyond Ferry's early suggestion of a shifted Rouse spectrum and the later Chömpff-Duiser supporting theory. However, the behavior of J_e^0 with molecular weight makes it clear that the terminal spectrum for long monodisperse chains is considerably narrower than the Rouse form. Explanations have been suggested, based on a presumed difference in the local velocity fields in entangled and nonentangled systems, but complete calculations starting from such ideas have not been made.

There is essentially no theoretical guidance about the structural variables that control the onset of entanglement effects, whether a critical contour length, a critical radius of gyration, or some critical combination of space-filling and mechanical properties of the chain. This problem presents many deep analytical difficulties. Computer simulation may provide some help here in developing intuition and suggesting some simple models for topological classification.

The relationship between chain entanglement and non-linear viscoelastic properties is somewhat equivocal. Shear rate dependence in viscosity is a rather general characteristic of long chain systems at all levels of concentration, including infinite dilution. It cannot therefore be solely an attribute of interchain interactions. Viscosity changes seem invariably to commence at shear rates which are of the order of the reciprocals of the longest relaxation times in the linear viscoelastic spectrum. The magnitude of these relaxation times is governed by intermolecular interactions in all but the most dilute range of concentrations. These properties are sufficient to eliminate potential intramolecular sources, such as internal viscosity and finite extensibility, as *primary* causes of shear rate dependence in concerned systems. The rapid increase in shear rate dependence with concentration suggests a close relationship with entanglement density in concentrated systems, and supports the idea that the degree of intermolecular coupling in such systems is reduced at high shear rates. The behavior of several other non-linear properties is also qualitatively consistent with the uncoupling hypothesis, but a fairly complete constitutive equation based on this idea and having real physical content is still lacking.

10. Acknowledgments

The author is grateful for the help of a number of people who have read and criticized various portions of this review. These include Dr. Gary Ver Strate, Prof. Arthur Lodge, Prof. Robert Simha, Prof. Michael Williams, Dr. Edward

Bagley, Prof. S. F. Edwards, Dr. Alfred Chömpff, and Dr. Toshiro Masuda. Unpublished data, pre-publication manuscripts and helpful comments were also provided by Prof. Guy Berry, Prof. G. Vinogradov, Prof. Shigeharu Onogi, Prof. Mitsuru Nagasawa, Dr. Neal Langley, Prof. Robert Bird, Dr. J. L. S. Wales, Dr. U. Daum, Prof. G. Allen, Prof. Donald Plazek, Prof. M. Kurata, Prof. J. Schurz, Dr. Nobu Nakajima, Prof. Robert Stratton, and Dr. Edward Collins. Prof. John Ferry's invitation to prepare the review, his technical comments during the course of the work, and his patience while awaiting its completion are much appreciated. Considerable help in data reduction and figure preparation was furnished by Ms. Kathryn Graessley. Finally, the author is grateful for the support of the National Science Foundation through grant GK-34362 during the time the review was being prepared.

11. Nomenclature

A	Matrix in the Rouse theory
a	Exponent in Mark-Houwink equation: $[\eta] = KM^a$.
a	Exponent in Fox-Allen $\eta_0 - M$ correlation.
a	Parameter of order unity in the continuous form of the Rouse spectrum.
b	Concentration exponent in J_e^0 and τ_m correlations.
b	$3 r_0^2/\langle r^2 \rangle$.
C	Total number of crosslinks (Part 7).
$C(t)$	Auto-correlation function for end-to-end vector.
C_1, C_2	Moduli in the Mooney-Rivlin elasticity equation.
C_{ij}, C_{ij}^{-1}	Components of the Cauchy-Green strain tensor and its inverse, with the current configuration as the reference configuration.
c	Polymer concentration in gm/ml.
c^*	Cornet critical concentration.
$c[\eta]$	Simha interaction parameter.
cM	Bueche interaction parameter.
D	Self-diffusion coefficient.
d	Power law exponent in $\eta \propto |\dot{\gamma}|^{-d}$.
E	Average number of entanglement points per molecule; $E = M/M_c - 1$.
E_0	Value of E at $\dot{\gamma} = 0$ (Part 8).
F	Viscosity structure factor.
F	Frictional force.
ΔF	Change in free energy with deformation.
f	Nominal tensile stress, force/initial cross sectional area.
f_i	Fraction of chain pairs belonging to topological class i.
$f(\beta)$	Reduced steady shear viscosity function $(\eta - \eta_s)/(\eta_0 - \eta_s)$.
G	Equilibrium shear modulus.
G^0	Instantaneous shear modulus in stress relaxation.
G_N^0	Shear modulus in the viscoelastic plateau region.
$G(t)$	Shear stress relaxation modulus.
$G'(\omega)$	Shear storage modulus.

$G''(\omega)$	Shear loss modulus.
$[G']_\omega$	Intrinsic storage modulus, $\lim\limits_{c\to 0}\left(\dfrac{G'(\omega,c)}{c}\right)$.
$[G'']_\omega$	Intrinsic loss modulus, $\lim\limits_{c\to 0}\left(\dfrac{G''(\omega,c)-\omega\eta_s}{c}\right)$.
g	Pair correlation function in molecular theory of liquids (Part 6).
g	Front factor in modulus equation from rubber elasticity theory (Part 7).
g_1, g_2	Fraction of configurations of free chains which are consistent with specified end-to-end coordinates (Part 7).
$g(\theta)$	Fractional reduction in entanglement density due to steady shear flow; $g(\theta) = E/E_0$ (Part 8).
$H(\tau)$	Relaxation time distribution.
h	Intramolecular hydrodynamic interaction parameter.
h^*	$h/N^{1/2}$.
$h(\theta)$	Fractional reduction in energy dissipation rate per molecule due to dis-entanglement in steady shear flow (Part 8).
$J(t)$	Shear creep compliance.
J_e^0	Steady state recoverable shear compliance.
J_{eR}	Reduced steady state compliance; $J_{eR} = J_e^0 c R T \eta_0^2 / M(\eta_0 - \eta_s)^2$.
$J^*(\xi)$	Steady state compliance for a monodisperse polymer of molecular weight ξ.
$J(\dot\gamma)$	$N_1/2\sigma^2$.
K	Unspecified proportionality constant.
K	Spring constant in the bead-spring models.
$K(s)$	Bueche entanglement slip function.
k	Boltzmann constant.
k'	Huggins constant.
$\mathbf{M}$	Forsman coupling matrix.
M	Molecular weight.
$M(\xi)$	Memory function in Lodge theory.
M_e	Molecular weight between entanglements in undiluted polymer.
M_c	Characteristic molecular weight from η_0 vs M behavior of undiluted polymers.
M_c'	Characteristic molecular weight from J_e^0 vs M behavior of undiluted polymers.
$\bar{M}_n$	Number-average molecular weight.
$\bar{M}_w$	Weight-average molecular weight.
$\bar{M}_z$	z-average molecular weight.
$\bar{M}_{z+1}$	$z+1$-average molecular weight.
M_x	General designation for characteristic molecular weights in the rheological behavior of undiluted polymers.
N	Number of sub-molecules in spring-bead models.
N	Number of primary molecules (Part 7).
N_a	Avogadro's number.
N_1	First normal stress function, $p_{11} - p_{22}$ at steady state in steady simple shear flow.
N_2	Second normal stress function, $p_{22} - p_{33}$ at steady state in steady simple shear flow.

n	Number of main chain atoms.
n_e	Number of main chain atoms between entanglements.
n_1, n_2	Moles of solvent and polymer respectively in solution thermodynamics (Part 2).
P	Degree of polymerization.
P_0	Unspecified isotropic pressure term in stress tensor p for incompressible materials.
p_{ij}	Component of stress tensor in rectangular coordinates.
R	Universal gas constant, kN_a.
R_g	Gel point radiation dose (Part. 2).
R_0	Stokes radius.
r	Position vector.
r_0	Contour length of polymer chain.
$\langle r^2 \rangle$	Mean-square end-to-end distance of polymer chain.
S	Mean radius of gyration, $\langle S^2 \rangle^{1/2}$.
ΔS	Entropy change with deformation.
s	Bueche slip factor.
T	Absolute temperature.
T_e	Langley entanglement trapping factor.
u	Unit vector.
V	Pervaded volume of polymer coil.
v	Velocity vector.
v	Speed, velocity magnitude.
v_0	Volume per main chain atom in undiluted polymer.
w_g	Gel fraction.
X	Structure parameter S^2/v_0 in Fox-Allen $\eta_0 - M$ correlation.
α	Coil expansion ratio (Parts 2 and 5).
α	Extension ratio in tensile deformation (Part 7).
β	Reduced shear rate, $(\eta_0 - \eta_s) M \dot{\gamma}/cRT$.
β'	Reduced frequency, $(\eta_0 - \eta_s) M \omega/cRT$.
β_0	Characteristic reduced shear rate locating the onset of shear rate dependence in the viscosity.
γ	Extent of simple shear deformation from rest state.
γ	Crosslink index, fraction of mers participating in crosslinks multiplied by DP_n of primary chains (Part 7).
γ_0	Instantaneously imposed shear deformation.
$\dot{\gamma}$	Shear rate.
$\dot{\gamma}_0$	Characteristic shear rate locating the onset of shear rate dependence in the viscosity.
$\delta(\ )$	Dirac delta function.
δ_{ij}	Kronecker delta function.
ε	Parameter characterizing the internal viscosity of chain molecules (Part 8).
ζ	Frictional coefficient.
ζ_0	Frictional coefficient per main chain atom.
ζ_e	Frictional coefficient associated with an entanglement junction.
η	Steady state shear viscosity, $\sigma(\dot{\gamma})/\dot{\gamma}$.
η'	Dynamic viscosity, $G'(\omega)/\omega$.

$\lvert\eta^*\rvert$	Absolute value of the complex viscosity $[G'(\omega)^2 + G''(\omega)^2]^{1/2}/\omega$.
η_0	Viscosity at zero shear rate.
η_s	Viscosity of solvent.
η_m	Viscosity of monomeric fluid in Eyring's theory.
η_c	Value of viscosity in undiluted polymer at $M = M_c$.
$[\eta]$	Intrinsic viscosity, $\displaystyle\lim_{c\to0}\left[\dfrac{\eta - \eta_s}{\eta_s c}\right]$.
$[\eta]_0$	Intrinsic viscosity at zero shear rate (Part 8).
θ	Theta condition for a polymer-solvent system.
θ	Argument $(\dot\gamma\tau_0/2)\,(\eta/\eta_0)$ in $g(\theta)$ and $h(\theta)$ functions of Graessley's theory (Part 8).
λ_i	Eigenvalues of transformation matrices.
$\lambda_1, \lambda_2, \lambda_3$	Intermolecular distances in Eyring's viscosity theory (Part 6).
ν	Chain concentration, molecules per unit volume.
ν	Concentration of elastically effective strands in crosslinked network (Part 7).
ν_e	Twice the concentration of entanglement junctions in a system prior to crosslinking (Part 7).
ν_0	Concentration of primary molecules prior to crosslinking (Part 7).
ν_c	Twice the concentration of crosslinks in a system (Part 7).
ϱ	Polymer density, mass/volume.
σ	Shear stress in simple shear flow.
τ	Relaxation time.
τ_0	Characteristic relaxation time associated with the onset of shear rate dependence in the viscosity.
τ_m	Characteristic "maximum" relaxation time determined from the terminal region of the viscoelastic spectrum.
τ_n	Number-average relaxation time of the terminal viscoelastic region, η_0/G_N^0.
τ_w	Weight-average relaxation time of the terminal viscoelastic region, $\eta_0 J_e^0$.
φ	Volume fraction of polymer.
Φ_∞	Flory constant, $[\eta]\,M/\langle r^2\rangle^{3/2}$ for linear flexible chains at high molecular weights.
χ	Polymer-solvent interaction coefficient (Part 2).
χ	Extinction angle in flow birefringence.
χ_0	Effective polymer-solvent interaction coefficient in determining chain dimensions in concentrated systems.
Ψ	Probability density distribution function for bead positions in the spring-bead molecular models.
Ψ_1	First normal stress coefficient, $N_1/\dot\gamma^2$.
Ψ_2	Second normal stress coefficient, $N_2/\dot\gamma^2$.
Ω	Number of distinguishable configurations.
ω	Frequency ω, rad/sec.
ω	Number of distinguishable configurations available to a free chain (Part 7).
ω_0	Characteristic frequency at which $\eta'(\omega)$ begins to depart from η_0.

12. References

1. Busse, W. F.: The physical properties of elastic colloids. J. Phys. Chem. **36**, 2862–2879 (1932).
2. Treloar, L. R. G.: Elastic recovery and plastic flow in raw rubber. Trans. Faraday Soc. **36**, 538–549 (1940).
3. Flory, P. J.: Network structure and the elastic properties of vulcanized rubber. Chem. Rev. **35**, 51–75 (1944).
4. Green, M. S., Tobolsky, A. V.: A new approach to the theory of relaxing polymers media. J. Chem. Phys. **14**, 80–92 (1946).
5. Buchdahl, R.: Rheology of thermoplastic materials. I. Polystyrene. J. Colloid Sci. **3**, 87–98 (1948).
6. Nielsen, L. E., Buchdahl, R.: Viscoelastic and photoelastic properties of polystyrene above its softening temperature. J. Colloid Sci. **5**, 282–294 (1950).
7. Bueché, F.: Viscosity, self-diffusion and allied effects in solid polymers. J. Chem. Phys. **20**, 1959–1964 (1952).
8. Bueche, F.: Viscosity of polymers in concentrated solutions. J. Chem. Phys. **25**, 599–600 (1956).
9. Fox, T. G., Flory, P. J.: Further studies on the melt viscosity of polyisobutylene. J. Phys. Colloid Chem. **55**, 221–234 (1951).
10. Fox, T. G., Loshaek, S.: Isothermal viscosity-molecular weight dependence for long chains. J. Appl. Phys. **26**, 1080–1082 (1955).
11. Ferry, J. D., Landel, R. F., Williams, M. L.: Extensions of the Rouse theory of viscoelastic properties to undiluted linear polymers. J. Appl. Phys. **26**, 359–362 (1955).
12. Lodge, A. S.: A network theory of flow birefringence and stress in concentrated polymer solutions. Trans. Faraday Soc. **52**, 120–130 (1956).
13. Yamamoto, M.: The visco-elastic properties of network structure. I. General formalism. J. Phys. Soc. Japan **11**, 413–421 (1956).
13a. Lodge, A. S.: The isotropy of gaussian molecular networks and the stress-birefringence relations for rubberlike materials cross-linked in stressed states. Kolloid-Z. **171**, 46–51 (1960).
13b. Bagley, E. B., West, D. C.: Chain entanglement in non-newtonian flow. J. Appl. Phys. **29**, 1511–1512 (1958).
14. Porter, R. S., Johnson, J. F.: The entanglement concept in polymer systems. Chem. Rev. **66**, 1–27 (1966).
15. Ferry, J. D.: Viscoelastic properties of polymers, 2nd ed. New York: Wiley 1970.
16. Berry, G. C., Fox, T. G.: The viscosity of polymers and their concentrated solutions. Adv. Polymer Sci. **5**, 261–357 (1968).
17. Dusek, K., Prins, W.: Structure and elasticity of non-crystalline polymer networks. Advan. Polymer Sci. **6**, 1–102 (1969).
18. Janeschitz-Kriegl, H.: Flow birefringence of elastico-viscous polymer systems. Advan. Polymer Sci. **6**, 170–318 (1969).
19. Berry, G. C., Casassa, E. F.: Thermodynamic and hydrodynamic behavior of dilute polymer solutions. J. Polymer Sci. Pt. D: Macromol. Rev. **4**, 1–66 (1970).
20. Fixman, M., Stockmayer, W. H.: Polymer conformation and dynamics in solution. Ann. Rev. Phys. Chem. **21**, 407–428 (1970).
21. Orofino, T. A.: Dilution-solution properties of polystyrene in θ-solvent media. II. An analysis of environmental effects. J. Chem. Phys. **45**, 4310–4315 (1966).
22. Flory, P. J.: Statistical mechanics of chain molecules. New York: Wiley-Interscience 1969.
23. Fixman, M., Peterson, J. M.: Dimensions of polymer molecules in concentrated solutions. J. Am. Chem. Soc. **86**, 3524–3529 (1964).
24. Yamakawa, H.: Modern theory of polymer solutions, pp. 214–215. New York: Harper and Row 1971.
25. Bluestone, S., Vold, M. J.: Monte-Carlo calculations of the dimensions of coiling type polymers in solutions of finite concentration. J. Polymer Sci. Pt. A **2**, 289–301 (1964).
26. Curro, J. G.: Computer simulation of semi-dilute polymeric systems. Paper presented at 1973 March Meeting of the American Physical Society, San Diego.

27. Flory, P. J.: Principles of polymer chemistry. Ithaca: Cornell University Press, 1953.
28. Krigbaum, W. R., Godwin, R. W.: Direct measurement of molecular dimensions in bulk polymers. J. Chem. Phys. **43**, 4523–4524 (1965).
29. Cotton, J. P., Farnoux, G., Jannink, G., Mons, J., Picot, C.: Observation de la conformation de la chaine polymerique dans la solide amorphe par diffusion de neutrons aux petits angles. Compt. Rend. Ser. C. **275**, 175–179 (1972).
30. Ballard, D. G. H., Wignall, G. D., Schelten, J.: Measurement of molecular dimensions of polystyrene chains in the bulk polymer by low angle neutron scattering. European Polymer J. **9**, 965–969 (1973).
31. Kirste, R. G., Kruse, W. A., Schelten, J.: Die Bestimmung des Trägheitsradius von Poly-methyl-methacrylat im Glaszustand durch Neutronenbeugung. Makromol. Chem. **162**, 299–303 (1972).
32. Schelten, J., Kruse, W. A., Kirste, R. G.: Über die Gestalt von Polymethylmethacrylat im Glaszustand. Kolloid-Z. u. Z. Polymere **251**, 919–921 (1973); see also J. Appl. Crystallography **7**, 188 (1974) and Polymer (London) (in press) by the same authors.
33. Bueche, F., Cashin, W. M., Debye, P.: The measurement of self-diffusion in solid polymers. J. Chem. Phys. **20**, 1956–1958 (1952).
34. Brandrup, J., Immergut, E. H. (Eds.): Polymer handbook. New York: Interscience 1966.
35. Bueche, F., Kinzig, B. J., Coven, C. J.: Polymer Chain Conformations in Bulk Polymers. Polymer Letters **3**, 399–402 (1965).
36. Smith, T. L., Frederick, J. E.: Ultimate tensile properties of elastomers. IV. Dependence of the failure envelope, maximum extensibility, and equilibrium stress-strain curve on network characteristics. J. Appl. Phys. **36**, 2996–3005 (1965).
37. Hoffman, M.: Zur Konformation von Molekülen in unverdünnten nichtkristallinen Polymeren. Makromol. Chem. **144**, 309–321 (1971).
38. Maron, S. H., Nakajima, N., Krieger, I. M.: Study of entanglement of polymers in solution by viscosity measurements. J. Polymer Sci. **37**, 1–18 (1959).
39. Maron, S. H.: A theory of the thermodynamic behavior of non-electrolyte solutions. J. Polymer Sci. **38**, 329–342 (1959).
40. Maron, S. H., Nakajima, N.: A theory of the thermodynamic behavior of non electrolyte solutions. II. Application to the system benzene-rubber. J. Polymer Sci. **40**, 59–71 (1959).
41. Maron, S. H., Nakajima, N.: A theory of the thermodynamic behavior of nonelectrolyte solutions. III. The osmotic pressure of polymer solutions. J. Polymer Sci. **42**, 327–340 (1966).
42. Maron, S. H., Nakajima, N.: A theory of the thermodynamic behavior of nonelectrolyte solutions. V. The scattering of light by polymer solutions. J. Polymer Sci. **47**, 157–168 (1960).
43. Aharoni, S. M.: Segmental density distribution of linear amorphous polymer molecules. J. Macromol. Sci.-Phys. B**7**, 73–103 (1973).
44. Vollmert, B., Stutz, H.: Zur Struktur von konzentrierten Polymerlösungen und Gelen. Angew. Makromol. Chem. **20**, 71–101 (1971). A description in English of these and later results by the same workers can be found on pp. 548–561 of Vollmert, B.: Polymer chemistry. Berlin-Heidelberg-New York: Springer 1973.
45. Duportail, G., Froelich, D., Weill, G.: Fluorescence quenching as a tool in the study of the distribution of local concentrations in polymer solutions. I. Polymers bearing quenching side groups. European Polymer J. **7**, 977–987 (1971).
46. Alberino, L. M., Graessley, W. W.: Cross-linking of polystyrene by high-energy radiation. II. Molecular weight changes in the pre-gel region. J. Phys. Chem. **72**, 4229–4235 (1968).
47. Budzol, M., Dole, M.: The radiation chemistry of polyethylene. XI. The molten state. J. Phys. Chem. **75**, 1671–1676 (1971).
48. Mandelkern, L.: Radiation chemistry of linear polyethylene, Vol. I. pp. 287–334. In: Dole, M. (Ed.): The radiation chemistry of macromolecules. New York: Academic Press 1972.
49. Flory, P.: Molecular configuration in bulk polymers. Pure Appl. Chem. (in press).
50. Kargin, V. A.: The role of structural phenomena in the formation of polymer properties. Pure Appl. Chem. **12**, 35–61 (1966).

51. Tager, A. A., Dreval, V. E., Lutsky, M. S., Vinogradov, G. V.: Rheological behavior of concentrated polystyrene solutions in good and poor solvents. J. Polymer Sci. Pt. C **23**, 181–192 (1968).

52. Tager, A. A., Dreval, V. E.: Non-newtonian viscosity of concentrated polymer solutions. Rheol. Acta **9**, 517–524 (1970).

53. Quadrat, O.: Viscometric properties of the concentrated polymer solutions. II. Dependence of the non-newtonian viscosity on the thermodynamic quality of the solvent and on its viscosity. Collection Czech. Chem. Commun. **37**, 3039–3041 (1972).

54. Debye, P., Chu, B., Woermann, D.: Viscosity of Critical Mixtures. J. Polymer Sci. Part A **1**, 249–254 (1963).

55. Ver Strate, G., Philippoff, W.: Phase separation in flowing polymer systems. Paper presented at 1973 March meeting of American Physical Society, San Diego.

56. Flory, P. J.: Statistical thermodynamics of semi-flexible chain molecules. Proc. Roy. Soc. (London) A**234**, 60–73 (1956).

57. Robertson, R. E.: Polymer order and polymer density. J. Phys. Chem. **69**, 1575–1578 (1965).

58. Yeh, G. S. Y., Geil, P. N.: Crystallization of polyethylene terephthalate from the glassy amorphous state. J. Macromol. Sci. B**1**, 235–249 (1967).

59. Carr, S. H., Geil, P. H., Baer, E.: The development of spherulites from structural units in glassy poly(bisphenol-A-carbonate). J. Macromol. Sci.-Phys. B**2**, 13–28 (1968).

60. Lin, W., Kramer, E. J.: Small-angle X-ray scattering from amorphous polycarbonate. J. Appl. Phys. **44**, 4288–4292 (1973).

61. Yeh, G. S. Y.: Morphology of amorphous polymers. CRC Critical Rev. Macromol. Sci. **1**, 173–213 (1972).

62. Yeh, G. S. Y.: Order in amorphous polystyrenes as revealed by electron diffraction and diffraction microscopy. J. Macromol. Sci.-Phys. B**6**, 451–464 (1972).

63. Yeh, G. S. Y.: A structural model for the amorphus state of polymers: folded-chain fringed micellar grain model. J. Macromol. Sci. Phys. B**6**, 465–478 (1972).

64. Hoeve, C. A. J., O'Brien, M. K.: Specific diluent effects on polymer chain dimensions. J. Polymer Sci. Part A **1**, 1947–1954 (1963).

65. Gent, A. N.: Stress-induced birefringence of swollen polymer networks. Macromolecules **2**, 262–265 (1969).

66. Tonelli, A. E.: On the existence of local segmental order in undiluted bulk polymers. J. Chem. Phys. **53**, 4339–4345 (1970).

67. Wellinghoff, S. T., Baer, E.: Crazing in ultra-thin films of atactic polystyrene. Unpublished manuscript.

68. Berens, A. R., Folt, V. L.: Resin particles as flow units in poly(vinyl chloride) melts. Trans. Soc. Rheol. **11**, 95–111 (1967).

68a. Collins, E. A., Krier, C. A.: Poly(vinyl chloride) melt rheology and flow activation energy. Trans. Soc. Rheol. **11**, 225–242 (1967).

69. Schreiber, H. P., Rudin, A., Bagley, E. B.: Separation of elastic and viscous effects in polymer melt extrusion. J. Appl. Polymer Sci. **9**, 887–892 (1965).

70. Ballman, R. L., Rademacher, L. E., Farnham, W. H.: Visualization of polystyrene melt flow. Presented at 1972 August U.S.-Japan joint seminar on polymer processing and rheology, Knoxville, Tenn. See J. Appl. Polymer Symp. **20** (1973).

71. Mooney, M., Wolstenholme, W. E.: The rheological unit in raw elastomers. J. Appl. Phys. **25**, 1098–1101 (1955).

72. Mooney, M.: Theory of the non-newtonian rheology of raw rubbers consisting of supermolecular rheological units. J. Appl. Phys. **27**, 691–696 (1956).

73. Prichard, J. H., Wissbrun, K. F.: Reversible melt flow rate increase in branched acetal polymers. J. Appl. Polymer Sci. **13**, 233–239 (1969).

74. Schreiber, H. P., Storey, S. H., Bagley, E. B.: Molecular fractionation in the flow of polymeric fluids. Trans. Soc. Rheol. **10**, 275–297 (1966).

75. Whitlock, L. R., Porter, R. S.: Experimental investigation of the concept of molecular migration within sheared polystyrene. J. Polymer Sci., Pt. A-2 **10**, 877–886 (1972).

76. Coleman, B. D., Noll, W.: Recent results in the continuum theory of viscoelastic fluids. Ann. N. Y. Acad. Sci. **89**, 672–714 (1961).

77. Coleman, B. D., Noll, W.: Simple fluids with fading memory, pp. 530–552. In: Proc. of the International Symposium on Second-Order Effects in Elasticity, Plasticity, and Fluid Dynamics, Haifa. New York: Pergamon Press 1962.

78. Lodge, A. S.: Elastic liquids. New York: Academic Press 1964.

79. Coleman, B. D., Markovitz, H., Noll, W.: Viscometric flows of non-newtonian fluids. Berlin-Heidelberg-New York: Springer 1966.

80. Coleman, B. D., Noll, W.: Foundations of linear viscoelasticity. Rev. Mod. Phys. **33**, 239–249 (1961).

81. Gross, B.: Mathematical structure of the theories of viscoelasticity. Paris: Hermann et Cie. 1953.

82. Tanner, R. I.: A correlation of normal stress data for polydisobutylene solutions. Trans. Soc. Rheol. **17**, 365–373 (1973).

83. Coleman, B. D., Markovitz, H.: Normal stress effects in second-order fluids. J. Appl. Phys. **35**, 1–9 (1964).

84. Stratton, R. A., Butcher, A. F.: Experimental determination of the relationships among various measures of fluid elasticity. J. Polymer Sci. Pt. A-2 **9**, 1703–1717 (1971).

85. Debye, P., Bueche, A. M.: Intrinsic viscosity, diffusion, and sedimentation rate of polymers in solution. J. Chem. Phys. **16**, 573–579 (1948).

86. Kramers, H. A.: The behavior of macromolecules in inhomogeneous flow. J. Chem. Phys. **14**, 415–424 (1946).

87. Kirkwood, J. G.: Collected works Macromolecules, Auer, P. L. (Ed.). New York: Gordon and Breach 1967.

88. Wang, M. C., Uhlenbeck, G. E.: On the theory of brownian motion. II. Rev. Mod. Phys. **17**, 323 (1945). In: Wax, N. (Ed.): Selected papers in noise and stochastic process. New York: Dover Publications.

89. Boyer, R. F.: The relation of transition temperature to chemical structure in high polymers. Rubber Chem. Tech. **36**, 1303–1421 (1963).

90. Clark, M. B., Zimm, B. H.: A linearized chain model for dielectric loss in polymers. ACS Polymer Preprints **12**, 116–120 (1971). See also: Tobolsky, A. V., DuPré, D. B.: Macromolecular relaxation in the damped torsional oscillator and statistical segment models. Advan. Polymer Sci. **6**, 103–127 (1969).

91. Iwata, I., Kurata, M.: Brownian motion of lattice-model polymer chains. J. Chem. Phys. **50**, 4008–4013 (1969).

92. Orwoll, R. A., Stockmayer, W. H.: Stochastic models for chain dynamics. Advan. Chem. Phys. **15**, 305–324 (1969).

93. Massa, D. J., Schrag, J. L., Ferry, J. D.: Dynamic viscoelastic properties of polystyrene in high-viscosity solvents. Extrapolation to infinite dilution and high-frequency behavior. Macromolecules **4**, 210–214 (1971).

94. Peterlin, A.: Molecular model of internal viscosity. Rheol. Acta **12**, 496–502 (1973).

95. Bazua, E. R., Williams, M. C.: A molecular formulation of the internal viscosity in polymer dynamics, and stress symmetry. J. Chem. Phys. **59**, 2858–2868 (1973).

96. Helfand, E.: Theory of the kinetics of conformational transitions in polymers. J. Chem. Phys. **54**, 4651–4661 (1971).

97. Rouse, P. E.: A theory of the linear viscoelastic properties of dilute solutions of coiling polymers. J. Chem. Phys. **21**, 1272–1280 (1953).

98. Bueche, F.: The viscoelastic properties of plastics. J. Chem. Phys. **22**, 603–609 (1954).

99. Chapter 10, Appendix B of Ref. (**27**).

100. Zimm, B. H.: Dynamics of polymer molecules in dilute solution: viscoelasticity, flow birefringence and dielectric loss. J. Chem. Phys. **24**, 269–278 (1956).

101. Lodge, A. S., Wu, Y.-J.: Constitutive equations for polymer solutions derived from the bead/spring model of Rouse and Zimm. Rheol. Acta **10**, 539–553 (1971).

102. Bird, R. B., Warner, H. R., Jr., Evans, D. C.: Kinetic theory and rheology of dumbbell suspensions with brownian motion. Advan. Polymer Sci. **8**, 1–90 (1971).

102a. Curtiss, C. F., Bird, R. B., Hassager, O.: Kinetic theory and rheology of dilute solutions of macromolecules. ACS Polymer Preprints **15**, 22–27 (1974).

103. Verdier, P. H., Stockmayer, W. H.: Monte Carlo calculations on the dynamics of polymers in dilute solution. J. Chem. Phys. **36**, 227–235 (1962). See also: Verdier, P. H.: Monte Carlo studies of lattice-model polymer chains. I. Correlation functions in the statistical-bead model. J. Chem. Phys. **45**, 2118–2121 (1966).
104. Tschoegl, N. W.: Influence of hydrodynamic interaction on the viscoelastic behavior of dilute polymer solutions in good solvents. J. Chem. Phys. **40**, 473–479 (1964).
105. Pyun, C. W., Fixman, M.: Intrinsic viscosity of polymer chains. J. Chem. Phys. **42**, 3838–3844 (1965).
106. Thurston, G. B., Morrison, J. D.: Eigenvalues and the intrinsic viscosity of short gaussian chains. Polymer (London) **10**, 421–438 (1969).
107. Lodge, A. S., Wu, Y.-J.: Exact relaxation times and dynamic functions for dilute polymer solutions from the bead/spring model of Rouse and Zimm, Report 16. Rheology Research Center, University of Wisconsin (July, 1972).
108. Williams, M. C.: Normal stresses in polymer solutions with remarks on the Zimm treatment. J. Chem. Phys. **42**, 2988–2989 (1965); errata J. Chem. Phys. **43**, 4542 (1965).
109. Ferry, J. D., Williams, M. L., Stern, D. M.: Slow relaxation mechanisms in concentrated polymer solutions. J. Phys. Chem. **58**, 987–992 (1954).
110. Verdier, P. H.: Monte Carlo studies of lattice model polymer chains. II. End-to-end length. J. Chem. Phys. **45**, 2122–2128 (1966).
111. Berry, G. C.: Thermodynamic and conformational properties of polystyrene. II. Intrinsic viscosity studies in dilute solutions of linear polystyrenes. J. Chem. Phys. **46**, 1338–1352 (1967).
112. Tanaka, H., Sakanishi, A., Kaneko, M., Furiuchi, J.: Dynamic viscoelastic properties of dilute polymer solutions. J. Polymer Sci. Pt. C **15**, 317–330 (1966).
113. Schrag, J. L., Johnson, R. M.: Application of the Birnboim multiple lumped resonator principle to viscoelastic measurements of dilute macromolecular solutions. Rev. Sci. Inst. **42**, 224–232 (1971).
114. Johnson, R. M., Schrag, J. L., Ferry, J. D.: Infinite-dilution viscoelastic properties of polystyrene in θ-solvents and good solvents. Polymer J. (Japan) **1**, 742–749 (1970).
115. Osaki, K., Mitsuda, Y., Johnson, R. M., Schrag, J. L., Ferry, J. D.: Infinite-dilution viscoelastic properties of linear and star-shaped polybutadienes. Macromol. **5**, 17–19 (1972).
116. Osaki, K., Schrag, J. L., Ferry, J. D.: Infinite-dilution viscoelastic properties of poly(α-methylstyrene). Applications of Zimm theory with exact eigenvalues. Macromolecules **5**, 144–147 (1972).
116a. Osaki, K.: A revised version of the integrodifferential equation in the Zimm theory for polymer solution dynamics. Macromolecules **5**, 141–144 (1972).
117. Osaki, K., Schrag, J. L.: Viscoelastic properties of polymer solutions in high-viscosity solvents and limiting high-frequency behavior. I. Polystyrene and poly(α-methylstyrene). Polymer J. (Japan) **2**, 541–549 (1971).
118. Thurston, G. B., Peterlin, A.: Influence of finite numbers of chain segments, hydrodynamic interaction, and internal viscosity on intrinsic birefringence and viscosity of polymer solutions in an oscillating laminar flow field. J. Chem. Phys. **46**, 4881–4885 (1967).
118a. Osaki, K.: Viscoelastic properties of dilute polymer solutions. Advan. Polymer Sci. **12**, 1–64 (1973).
119. Simha, R., Somcynsky, T.: The viscosity of concentrated spherical suspensions. J. Colloid Sci. **20**, 278–281 (1965).
120. Batchelor, G. K., Green, J. T.: The determination of the bulk stress in a suspension of spherical particles to order c^2. J. Fluid Mech. **56**, 401–427 (1972).
121. Gandhi, K. S., Williams, M. C.: Solvent effects on the viscosity of moderately concentrated polymer solutions. J. Polymer Sci. Pt. C **35**, 211–234 (1971).
122. Frederick, J. E., Tschoegl, N. W., Ferry, J. D.: Dynamic mechanical properties of dilute polystyrene solutions; dependence on molecular weight, concentration, and solvent. J. Phys. Chem. **68**, 1974–1982 (1964).
123. Holmes, L. A., Kusamizu, K., Osaki, K., Ferry, J. D.: Dynamic mechanical properties of moderately concentrated polystyrene solutions. J. Polymer Sci. Pt. A-2 **9**, 2009–2021 (1971).

124. Onogi,S., Masuda,T., Kitagawa,K.: Rheological properties of anionic polystyrenes. I. Dynamic viscoelasticity of narrow-distribution polystyrenes. Macromolecules **3**, 109–116 (1970).

125. Frisch,H.L., Simha,R.: The viscosity of colloidal suspensions and macromolecular solutions, Vol. 1, Chapter 14. In: Eirich,F.R. (Ed.): Rheology. New York: Academic Press 1956.

126. Simha,R., Zakin,J.L.: Compression of flexible chain molecules in solution. J. Chem. Phys. **33**, 1791–1793 (1960).

127. Onogi,S., Kobayashi,Y., Kojima,Y., Taniguchi,Y.: Non-Newtonian flow of concentrated solutions of high polymers. J. Appl. Polymer Sci. **7**, 847–859 (1963).

128. Cornet,C.F.: The determination of unperturbed dimensions of polymer molecules by viscometry of moderately concentrated solutions. Polymer (London) **6**, 373–384 (1965).

129. Utracki,L., Simha,R.: Corresponding states relations for the viscosity of moderately concentrated polymer solutions. J. Polymer Sci. Pt. A **1**, 1089–1098 (1963).

130. Simha,R., Utracki,L.A.: The viscosity of concentrated polymer solutions: corresponding states principles. Rheol. Acta **12**, 455–462 (1973).

131. Simha,R., Chan,F.S.: Corresponding states relations for the newtonian viscosity of concentrated polymer solutions. Temperature dependence. J. Phys. Chem. **75**, 256–267 (1971).

132. Onogi,S., Kimura,S., Kato,T., Masuda,T., Miyanaga,N.: Effects of molecular weight and concentration on flow properties of concentrated polymer solutions. J. Polymer Sci. Pt. C **15**, 381–406 (1966).

133. Onogi,S., Masuda,T., Miyanaga,N., Kimura,Y.: Dependence of viscosity of concentrated polymer solutions upon molecular weight and concentration. J. Polymer Sci. Pt. A-2 **5**, 899–913 (1967).

134. Gandhi,K.S., Williams,M.C.: Effect of solvent character on polymer entanglements. J. Appl. Polymer Sci. **16**, 2721–2725 (1972).

135. Dreval,V.Ye., Tager,A.A., Fomina,A.S.: Concentrated solutions of polymers. IV. Viscosity of solutions of polystyrene in various solvents. Polymer Sci. USSR **5**, 495–507 (1964) [Vysokomolekul. Soyedin. **5**, 1404 (1963)].

136. Ferry,J.D., Foster,E.L., Browning,G.V., Sawyer,W.M.: Viscosities of concentrated polyvinyl acetate solutions in various solvents. J. Colloid Sci. **6**, 377–388 (1951).

137. Simha,R., Zakin,J.L.: Solution viscosities of linear flexible high polymers. J. Colloid Sci. **17**, 270–287 (1962).

138. Quadrat,O., Podnecka,J.: Influence of the thermodynamic quality of a solvent upon the viscosity of moderately concentrated polystyrene solutions. Collection Czech. Chem. Commun. **37**, 2402–2409 (1972).

139. Johnson,M.F., Evans,W.W., Jordan,I., Ferry,J.D.: Viscosities of concentrated polymer solutions. II. Polyisobutylene. J. Colloid Sci. **7**, 498–510 (1952).

140. Ferry,J.D., Grandine,L.D.,Jr., Udy,D.C.: Viscosities of concentrated polymer solutions. III. Polystyrene and styrene-maleic acid copolymer. J. Colloid Sci. **8**, 529–539 (1953).

140a. Toms,B.A., Strawbridge,D.J.: Elastic and viscous properties of dilute solutions of polymethyl methacrylate in organic liquids. Trans. Faraday Soc. **49**, 1225–1232 (1953).

141. Dreval,V.Y2., Malkin,A.Ya., Vinogradov,G.V.: Effect of the solvent nature on the rheological properties of concentrated solutions of various polymers. European Polymer J. **9**, 85–99 (1973).

142. Masuda,T., Toda,N., Aoto,Y., Onogi,S.: Viscoelastic properties of concentrated solutions of poly(methyl methacrylate) in diethyl phthalate. Polymer J. (Japan) **3**, 315–321 (1972).

143. Klein,J., Woernle,R.: Zum Einfluß der Polymer-Lösungsmittel-Wechselwirkung auf das Fließverhalten mäßig konzentrierter Polymerlösungen. Kolloid-Z. Z. Polymere **237**, 209–219 (1970).

144. Borchard,W., Pyrlik,M., Rehage,G.: Association phenomena of PMMA in solution and gels. Makromol. Chem. **145**, 169–188 (1971).

145. Rotne,J., Prager,S.: Variational treatment of hydrodynamic interaction in polymers. J. Chem. Phys. **50**, 4831–4837 (1969).

146. Masuda,T., Kitagawa,K., Onogi,S.: Viscoelastic properties of poly (methyl methacrylates) prepared by anionic polymerization. Polymer J. (Japan) 1, 418–424 (1970).
147. Onogi,S., Masuda,T., Ibaragi,T.: Rheological properties of polymethyl methacrylate and polyvinyl acetate in the molten state. Kolloid-Z. Z. Polymere 222, 110–124 (1968).
148. Ninomiya,K., Ferry,J.D., Oyanagi,Y.: Viscoelastic properties of polyvinyl acetates. II. Creep studies of blends. J. Phys. Chem. 67, 2297–2308 (1963).
149. Peterlin,A.: Gradient and time dependence of viscosity of polymer solutions in very viscous solvents. J. Lubrication Tech. 90, 571–576 (1968).
150. Lodge,A.S.: Rheological properties of concentrated polymer solutions. I. Growth of pressure fluctuations during prolonged shear flow. Polymer (London) 2, 195–201 (1961).
151. Philippoff,W.: Studies of flow birefringence of polystyrene solutions. Trans. Soc. Rheol. 7, 45–59 (1963).
152. Osaki,K., Einaga,Y.: Viscoelastic properties of concentrated polymer solutions. Prog. Polymer Sci. Japan 1, 321–375 (1971).
153. Odani,H., Nemoto,N., Kurata,M.: The viscoelastic properties of undiluted linear polymers of narrow molecular weight distribution in the terminal zone. Bull. Inst. Chem. Res., Kyoto U. 50, 117–133 (1972).
154. Allen,V.R., Fox,T.G.: Viscosity-molecular weight dependence for short chain polystyrenes. J. Chem. Phys. 41, 337–343 (1964).
155. Graessley,W.W., Hazleton,R.L., Lindeman,L.R.: The shear-rate dependence of viscosity in concentrated solutions of narrow-distribution polystyrene. Trans. Soc. Rheol. 11, 267–285 (1967).
156. Gupta,D., Forsman,W.C.: Newtonian viscosity-molecular weight relationship for concentrated solutions of monodisperse polystyrene. Macromolecules 2, 304–306 (1969).
157. Berry,G.C.: The viscosity of polymer-diluent mixtures. J. Phys. Chem. 70, 1194–1198 (1966).
158. Fetters,L.J.: Determination of the intermolecular entanglement coupling spacings in polyisoprene by viscosity measurements. J. Res. Natl. Bur. Std. 69A, 33–37 (1965).
159. Schurz,J., Hochberger,H.: Untersuchungen über den Lösungszustand von mäßig konzentrierten Polyisobutylen-Lösungen. Makromol. Chem. 96, 141–149 (1966).
160. Cross,M.M.: Viscosity, molecular weight, and chain entanglement. Polymer (London) 11, 238–244 (1970).
161. Fujimoto,T., Ozaki,N., Nagasawa,M.: Stress relaxation of monodisperse poly-α-methylstyrene. J. Polymer Sci. Pt. A-2 6, 129–140 (1968).
162. Tobolsky,A.V.: Properties and structure of polymers. New York: Wiley 1960.
163. Plazek,D.J., O'Rourke,V.M.: Viscoelastic behavior of low molecular weight polystyrene. J. Polymer Sci. Pt. A-2 9, 209–243 (1971).
164. Barlow,A.J., Day,M., Harrison,G., Lamb,J., Subramanian,S.: Viscoelastic relaxation in a series of polyethylacrylates and poly-n-butylacrylates. Proc. Roy. Soc. A 309, 497–525 (1969).
165. Knoff,W.F., Hopkins,I.L., Tobolsky,A.V.: Studies on the stress relaxation of polystyrenes in the rubbery flow region. II. Macromolecules 4, 750–754 (1971).
166. Nemoto,N., Moriwaki,M., Odani,H., Kurata,M.: Shear creep studies of narrow-distribution poly(cis-isoprene). Macromolecules 4, 215–219 (1971).
167. Nemoto,N., Ogawa,T., Odani,H., Kurata,M.: Shear creep studies of narrow-distribution poly(cis-isoprene). III. Concentrated solutions. Macromolecules 5, 641–644 (1972).
168. Oser,H., Marvin,R.S.: Effect of molecular weight on viscoelastic properties of polymers as predicted by a molecular theory. J. Res. Natl. Bur. Std. 67B, 87–90 (1963).
169. Andrews,R.D., Tobolsky,A.V.: Elastoviscous properties of polyisobutylene. IV. Relaxation time spectrum and calculation of bulk viscosity. J. Polymer Sci. 7, 221–242 (1951).
170. De Witt,T.W., Markovitz,H., Padden,F.J., Jr., Zapas,L.J.: Concentration dependence of the rheological behavior of the polyisobutylene — decalin system. J. Colloid. Sci. 10, 174–188 (1955).

171. Tobolsky, A. V., Aklonis, J. J., Akovali, G.: Viscoelastic properties of monodisperse polystyrene. J. Chem. Phys. **42**, 723–728 (1965).
172. Kotaka, T., Kurata, M., Tamura, M.: Non-Newtonian flow and normal stress phenomena in solutions of polystyrene in toluene. Rheol. Acta **2**, 179–186 (1962).
173. Kotaka, T., Osaki, K.: Normal stresses, non-Newtonian flow, and dynamic mechanical behavior of polymer solutions. J. Polymer Sci. Pt. C **15**, 453–479 (1966).
174. Tung, L. H., Runyon, J. R.: Molecular weight distribution of standard polystyrene samples by GPC and by sedimentation velocity analysis. J. Appl. Polymer Sci. **17**, 1589–1596 (1973).
175. Holmes, L. A., Ferry, J. D.: Dependence of the steady-state compliance on concentration and molecular weight in polymer solutions. J. Polymer Sci. Pt. C **23**, 291–299 (1968).
176. Kusamizu, S., Holmes, L. A., Moore, A. A., Ferry, J. D.: The steady-state compliance of polymer solutions. Trans, Soc. Rheol. **12**, 559–571 (1968).
177. Ashare, E.: Rheological properties of monodisperse polystyrene solutions, Ph. D. Thesis, University of Wisconsin (1968). Trans. Soc. Rheol. **12**, 535–557 (1968).
178. Graessley, W. W., Segal, L.: Flow behavior of polystyrene systems in steady shearing flow. Macromolecules **2**, 49–57 (1969).
179. Daum, U.: (Private communication), see Ref. (*190*).
180. Wales, J. L. S.: (Private communication), see Ref. (*190*).
181. Prest, W. M., Jr.: Viscoelastic properties of blends of entangled polymers, J. Polymer Sci: Pt. A-2 **8**, 1897–1908 (1970).
182. Einaga, Y., Osaki, K., Kurata, M., Tamura, M.: Creep behavior of polymer solutions. II. Steady-shear compliance of concentrated polystyrene solutions. Macromolecules **4**, 87–92 (1971).
183. Tschoegl, N. W.: Stress relaxation and creep in dilute polymer solutions. J. Chem. Phys. **44**, 2331–2334 (1966).
184. Akers, L. C., Williams, M. C.: Oscillatory normal stresses in dilute polymer solutions. J. Chem. Phys. **51**, 3834–3841 (1969).
185. Stratton, R. A., Butcher, A. F.: Experimental determination of the relationships among various measures of fluid elasticity. J. Polymer Sci.: Pt A-2 **9**, 1703–1717 (1971).
186. Osaki, K., Einaga, Y., Kurata, M., Tamura, M.: Creep behavior of polymer solutions. I. A new kind of apparatus for creep and creep recovery. Macromolecules **4**, 82–87 (1971).
187. Sakai, M., Fujimoto, T., Nagasawa, M.: Steady flow properties of monodisperse polymer solutions. Molecular weight and polymer concentration dependences of steady shear compliances at zero and finite shear rates. Macromolecules **5**, 786–792 (1972).
188. Akovali, G.: Viscoelastic properties of polystyrene. J. Polymer Sci. Pt. A-2 **5**, 875–889 (1967).
189. Mieras, H. J. M. A., van Rijn, C. F. H.: Elastic behavior of some polymer melts. Nature **218**, 865–866 (1968).
190. Daum, U., Wales, J. L. S.: Critical concentrations in polystyrene solutions. Polymer Letters **7**, 459–462 (1969).
191. Nemoto, N.: Viscoelastic properties of narrow-distribution polymers. II. Tensile creep studies of polystyrene. Polymer J. (Japan) **1**, 485–492 (1970).
192. Mills, N. J., Nevin, A.: Oscillatory shear measurements on polystyrene melts in the terminal region. J. Polymer Sci. Pt. A-2 **9**, 267–281 (1971).
193. Crawley, R.: Unpublished results in M. S. thesis. Northwestern University, 1973.
194. Zosel, A.: Der Einfluß von Molekulargewicht und Moleculargewichtsverteilung auf die viskoelastischen Eigenschaften von Polystyrolschmelzen. Rheol. Acta **10**, 215–224 (1971).
195. Uy, W. C., Graessley, W. W.: Viscosity and normal stresses in poly(vinyl acetate) systems. Macromolecules **4**, 458–463 (1971).
196. Nemoto, N., Odani, H., Kurata, M.: Shear creep studies of narrow distribution poly(cis-isoprene). II. Extension to low molecular weights. Macromolecules **5**, 531–535 (1972).
197. Vinogradov, G. V., Malkin, A. Ya., Yanovskii, Yu. G., Borisenkova, E. K., Yarlykov, B. V., Berezhnaya, G. V.: Viscoelastic properties and flow of narrow distribution polybutadienes and polyisoprenes. J. Polymer Sci. Pt. A-2 **10**, 1061–1084 (1972).

198. Endo,H., Fujimoto,T., Nagasawa,M.: Normal stress and shear stress in a viscoelastic liquid under steady state flow: Effect of molecular weight heterogeneity. J. Polymer Sci. Pt. A-2 **9**, 345–362 (1971).
199. Sakai,M., Fujimoto,T., Nagasawa,M.: Steady flow properties of monodisperse polymer solutions. Molecular weight and polymer concentration dependences of steady shear compliances at zero and finite shear rates. Macromolecules **5**, 786–792 (1972).
200. Odani,H., Nemoto,N., Kitamura,S., Kurata,M.: Viscoelastic properties of mono-disperse polymers. I. Tensile creep studies of poly(α-methylstyrene). Polymer J. (Japan) **1**, 356–364 (1970).
201. Osaki,K., Sakato,K., Fukatsu,M., Kurata,M., Matusita,K., Tamura,M.: Normal stress effect in dilute polymer solutions. III. Monodisperse poly-α-methylstyrene in chlorinated biphenyl. J. Phys. Chem. **74**, 1752–1756 (1970).
202. Valentine,R.H.: Viscoelastic properties of linear and cross-linked polybutadienes. Ph. D. thesis, University of Wisconsin, 1967.
203. Kraus,G.: Private communication.
204. Mills,N.J.: The rheological properties and molecular weight distribution of poly-dimethylsiloxane. European Polymer J. **5**, 675–695 (1969).
205. Ninomiya,K., Ferry,J.D.: Viscoelastic properties of polyvinyl acetates. I. Creep studies of fractions. J. Phys. Chem. **76**, 2292–2296 (1963).
206. Oyanagi,Y., Ferry,J.D.: Viscoelastic properties of polyvinyl acetates. IV. Creep studies of plasticized fractions. J. Colloid Sci. **21**, 547–559 (1966).
207. Mieras,H.J.M.A.: Elastic or normal-stress behavior of monodisperse polystyrene melts or solutions. Paper presented at the conference Advances in Rheology, Glasgow, September 1969.
208. Prest,W.M., Jr., Porter,R.S., O'Reilly,M.M.: Non-Newtonian flow and the steady-state shear compliance. J. Appl. Polymer Sci. **14**, 2697–2706 (1970).
209. Masuda,T., Kitagawa,K., Inoue,T., Onogi,S.: Rheological properties of anionic poly-styrenes. II. Dynamic viscoelasticity of blends of narrow-distribution polystyrenes. Macromolecules **3**, 116–125 (1970).
210. Saeda,S., Yotsuyanagi,J., Yamaguchi,K.: Relation between melt flow properties and molecular weight in polyethylene. J. Appl. Polymer Sci. **15**, 277–297 (1971).
210a. Locati,G., Gargani,L.: Dependence of zero-shear viscosity on molecular weight distribution. J. Polymer Sci., Polymer Letters **11**, 95–101 (1973).
211. Pennline,H.W., Graessley,W.W.: Flow properties of polymethyl methacrylate solutions. Paper presented at American Physical Society Meeting, San Diego, March 1973.
212. Graessley,W.W.: Linear viscoelasticity in entangling polymer systems. J. Chem. Phys. **54**, 5143–5157 (1971).
213. Leaderman,H., Smith,R.G., Williams,L.C.: Rheology of polyisobutylene. III. Elastic recovery, non-Newtonian flow, and molecular weight distribution. J. Polymer Sci. **36**, 233–257 (1959).
214. Ninomiya,K., Ferry,J.D.: Phenomenological relation for viscoelastic properties of polymer diluent systems; the relative dissipation index N_{21}. J. Macromol. Sci.-Phys. B **3**, 237–258 (1969).
215. Bogue,D.C., Masuda,T., Einaga,Y., Onogi,S.: A constitutive model for molecular weight and concentration effects in polymer blends. Polymer J. (Japan) **1**, 563–572 (1970).
216. Chikahisa,Y.: A theory on the relationship between viscosity and molecular weight in bulk polymers. J. Phys. Soc. Japan **19**, 92–100 (1964).
217. Williams,M.C.: Stresses in concentrated polymer solutions. Part I. Shear dependence of viscosity. A.I.Ch.E. J. **12**, 1064–1070 (1966).
218. Williams,M.C.: Concentrated polymer solutions. Part II. Dependence of viscosity and relaxation time on concentration and molecular weight. A.I.Ch.E. J. **13**, 534–539 (1967).
219. Williams,M.C.: Concentrated polymer solutions. Part III. Normal stresses in simple shear flow. A.I.Ch.E. J. **13**, 955–961 (1967).
220. Fixman,M.: Dynamics of polymer chains. J. Chem. Phys. **42**, 3831–3837 (1965).
221. Middleman,S.: The flow of high polymers. New York: Interscience 1968.

222. Lodge, A. S.: Constitutive equations from molecular network theories for polymer solutions. Rheol. Acta 7, 379–392 (1968).
223. Flory, P. J.: Elasticity of polymer networks cross-linked in state of strain. Trans. Faraday Soc. 56, 722–743 (1960).
224. Bueche, F.: Viscosity of polymers in concentrated solution, J. Chem. Phys. 25, 599–600 (1956). See also: Bueche, F.: Viscosity of molten branched polymers and their concentrated solutions. J. Chem. Phys. 40, 484–487 (1964).
225. De Gennes, P. G.: Reptation of a polymer chain in the presence of fixed obstacles. J. Chem. Phys. 55, 572–579 (1971).
226. Eyring, H., Ree, T., Hirai, N.: The viscosity of high polymers — Random walk of a group of connected segments. Proc. Natl. Acad. Sci. 44, 1213–1217 (1959).
227. Graessley, W. W.: Viscosity of entangling polydisperse polymers. J. Chem. Phys. 47, 1942–1953 (1967).
228. Aharoni, S. M.: On some rheological phenomena of amorphous polymers. J. Appl. Polymer Sci. 17, 1507–1518 (1973).
229. Ziabicki, A., Takserman-Krozer, R.: General dynamic theory of macromolecular networks. I. Definitions and classification. J. Polymer Sci. Part A-2 7, 2005–2018 (1969).
230. Takserman-Krozer, R., Ziabicki, A.: General dynamic theory of macromolecular networks. II. Dynamics of network deformation. J. Polymer Sci. Part A-2 8, 321–332 (1970).
231. Ziabicki, A.: Molecular rheology of polymer systems. Pure Appl. Chem. 26, 481–497 (1971).
232. Chömpff, A. J., Duiser, J. A.: Viscoelasticity in networks consisting of crosslinked or entangled macromolecules. I. Normal modes and mechanical spectra. J. Chem. Phys. 45, 1505–1514 (1966).
233. Duiser, J. A., Staverman, A. J.: On the theory of rubber elasticity. In: Prins, J. A. (Ed.): Physics of non-crystalline solids, pp. 376–387. Amsterdam: North Holland Publ. 1965.
234. Chömpff, A. J.: Normal modes of branched polymers. I. Simple ring and star-shaped molecules. J. Chem. Phys. 53, 1566–1576 (1970).
235. Chömpff, A. J., Prins, W.: Viscoelasticity of networks consisting of crosslinked or entangled macromolecules. II. Verification of the theory for entanglement networks. J. Chem. Phys. 48, 235–243 (1968).
236. Edwards, S. F., Grant, J. W. V.: Effect of entanglement on diffusion in a polymer melt. J. Phys. A: Math. Nucl. Gen. 6, 1169–1185 (1973).
237. Edwards, S. F., Grant, J. W. V.: The effect of entanglement on the viscosity of a polymer melt. J. Phys. A, Math. Nucl. Gen. 6, 1186–1195 (1973).
238. Forsman, W. C., Grand, H. S.: Theory of entanglement effects in linear viscoelastic behavior of polymer solutions and melts. I. Symmetry considerations. Macromolecules 5, 289–293 (1972).
239. Wolkowicz, R. I., Forsman, W. C.: Entanglement in concentrated solutions of polystyrene with narrow distributions of molecular weight. Macromolecules 4, 184–192 (1971).
239a. Thirion, P.: Empirical extension of the molecular theory of Rouse to the viscoelasticity of bulk polymers. J. Polymer Sci., Polymer Letters Ed. 11, 673–678 (1973).
240. Hoffman, M.: Das Grundgesetz für die mechanische Relaxation und das Fließen unvernetzter Fadenmoleküle. Rheol. Acta 6, 92–100 (1967).
241. Graessley, W. W.: Unpublished calculations.
242. Vinogradov, G. V., Pokrovsky, V. H., Yanovsky, Yu. G.: Theory of viscoelastic behavior of linear polymers in unimolecular approximation and its experimental verification. Rheol. Acta 11, 258–274 (1972).
243. Hayashi, S.: Theory of viscoelasticity in temporarily crosslinked polymers. IV. Relaxation spectrum and molecular weight dependence of viscosity. J. Phys. Soc. Japan 19, 2306–2312 (1964).
244. Hayashi, S.: Concentration dependence of rheological properties in concentrated polymer solutions. In: Onogi, S. (Ed.): Proceedings of the Fifth International Congress of Rheology, Vol. 4, pp. 179–190. Baltimore: University Park Press 1969.

245. Fox, T.G., Allen, V.R.: Dependence of the zero-shear melt viscosity and the related friction coefficient and critical chain length on measurable characteristics of chain polymers. J. Chem. Phys. **41**, 344–352 (1964).

246. Hoffmann, M.: Der Einfluß von Verhakungen auf die mechanische Relaxation und das Fließen der amorphen Polymeren. Rheol. Acta **6**, 377–390 (1967).

247. Tonelli, A.E.: A molecular approach to chain entanglement in linear polymers. J. Polymer Sci. Part A-2 **8**, 625–635 (1970).

248. Pechhold, W.: Molekülbewegung in Polymeren. I. Teil: Konzept einer Festkörperphysik makromolekularer Stoffe. Kolloid-Z. Z. Polymere **228**, 1–38 (1968).

249. Bueche, F.: Melt viscosity of polymers: effect of polydispersity. J. Polymer Sci. **43**, 527–530 (1960).

250. Prest, W.M., Jr., Porter, R.S.: The effect of high-molecular-weight components on the viscoelastic properties of polystyrene. Polymer J. (Japan) **4**, 154–162 (1973).

251. Case, L.C., Wargin, R.V.: Elastomer behavior. IV. The loop structure of elastomer networks. Makromol. Chem. **77**, 172–184 (1964).

252. Alfrey, T., Jr., Lloyd, W.G.: Network polymers. I. Theoretical remarks. J. Polymer Sci. **62**, 159–165 (1962).

253. Moore, C.G., Watson, W.F.: Determination of degree of crosslinking in natural rubber vulcanizates. Part II. J. Polymer Sci. **19**, 237–254 (1956).

254. Bueche, A.M.: An investigation of the theory of rubber elasticity using irradiated poly(dimethylsiloxanes). J. Polymer Sci. **19**, 297–306 (1956).

255. Langley, N.R.: Elastically effective strand density in polymer networks. Macromolecules **1**, 348–352 (1968).

256. Treloar, L.R.G.: The physics of rubber elasticity, 2nd ed. London: Oxford University Press 1958.

256a. Smith, K.J., Jr.: Theories of chain coiling, elasticity, and viscoelasticity. In: Jenkins, A.D. (Ed.): Polymer science, Chapter 5. Amsterdam: North-Holland Publ. Co. 1972.

257. Mark, J.E.: An experimental comparison of the theories of elasticity of polymer networks. J. Am. Chem. Soc. **92**, 7252–7257 (1970).

258. Rivlin, R.S.: Large elastic deformations. In: Eirich, F.R. (Ed.): Rheology, Vol. 1, pp. 351–385. New York: Academic Press 1956.

259. Imai, S., Gordon, M.: Rubber elasticity. J. Chem. Phys. **50**, 3889–3903 (1969); also note correction regarding g on p. 9 of Ref. (*284*).

260. Edwards, S.F.: The statistical mechanics of rubbers. In: Chömpff, A.J., Newman, S. (Eds.): Polymer networks, structure, and mechanical properties, pp. 83–110. New York: Plenum Press 1971.

261. Freed, K.F.: Statistical mechanics of systems with internal constraints: rubber elasticity. J. Chem. Phys. **55**, 5588–5599 (1971).

262. Eichinger, B.E.: Elasticity theory. I. Distribution functions for perfect phantom networks. Macromolecules **5**, 496–505 (1972).

263. Guth, E.: Recent developments in the statistical mechanics of polymers. J. Polymer Sci. Pt. C **31**, 267–274 (1970).

264. Labana, S.S., Newman, S., Chömpff, A.J.: Chemical effects on the ultimate properties of polymer networks in the glassy state, pp. 453–477. In: Polymer networks, structure and mechanical properties. See Ref. (*260*).

265. Dusek, K.: Inhomogeneities induced by crosslinking in the course of crosslinking copolymerization, pp. 245–260. In: Polymer networks, structure, and mechanical properties. See Ref. (*260*).

266. Goebel, J.C., Tobolsky, A.V.: Volume changes accompanying rubber extension. Macromolecules **4**, 208–209 (1971).

267. Krigbaum, W.R., Roe, R.-J.: Survey of theory of rubberlike elasticity. Rubber Chem. Tech. **38**, 1039–1067 (1965).

268. Valanis, K.C., Landel, R.F.: The strain-energy function of a hyperelastic material in terms of extension ratios. J. Appl. Phys. **38**, 2997–3002 (1967).

269. Thirion, P., Chasset, R.: Les rapports entre l'elasticité et la structure des réseaux macromoléculaires: Esquisse d'une nouvelle approche. R.G.C.P. **45**, 859–866 (1968).

269a. Yu, C.U., Mark, J.E.: Evaluation of the theories of rubber-like elasticity using swollen networks crosslinked at different degrees of dilution. Macromolecules **6**, 751–754 (1973).

270. Allen, G., Kirkham, M. J., Padget, J., Price, C.: Thermodynamics of rubber elasticity at constant volume. Trans Faraday Soc. **67**, 1278–1292 (1971).

271. Gumbrell, S. M., Mullins, L., Rivlin, R. S.: Departures of the elastic behavior of rubbers in simple extensions from the kinetic theory. Trans. Faraday Soc. **49**, 1495–1506 (1953).

272. Van der Hoff, B. M. E.: The stress-strain relation of swollen rubbers. Polymer (London) **6**, 397–399 (1965).

273. Mark, J. E., Flory, P. J.: Stress-strain isotherms for poly-(dimethylsiloxane) networks. J. Appl. Phys. **37**, 4635–4639 (1966).

274. Price, C., Allen, G., de Candia, F., Kirkham, M. C., Subramaniam, A.: Stress-strain behavior of natural rubber vulcanized in the swollen state. Polymer (London) **11**, 486–491 (1970).

275. Johnson, R. M., Mark, J. E.: Properties of poly(dimethylsiloxane) networks prepared in solution, and their use in evaluating the theories of rubberlike elasticity. Macromolecules **5**, 41–45 (1972).

276. Gent, A. N., Rivlin, R. S.: Experiments on the mechanics of rubber. II. The torsion, inflation, and extension of a tube. Proc. Phys. Soc. (Pt. B) **65**, 487–501 (1952).

277. Van der Hoff, B. M. E., Buckler, E. J.: Transient changes in topology and energy on extension of polybutadiene networks. J. Macromol. Sci. (Chem.) A **1**, 747–788 (1967).

278. Schaefgen, J. R., Flory, P. J.: Multilinked polyamides. J. Am. Chem. Soc. **72**, 689–701 (1950).

279. Kraus, G., Moczygemba, G. A.: Chain entanglements and elastic behavior in poly-butadiene networks. J. Polymer Sci. Part A **2**, 277–288 (1964).

280. Froelich, D., Crawford, D., Rozek, T., Prins, W.: Ideal network behavior of anionically prepared polystyrene gels. Macromolecules **5**, 100–102 (1972).

280a. Allen, G., Holmes, P. A., Walsh, D. J.: To be published in Disc. Faraday Soc. See also: Walsh, D. J., Allen, G., Ballard, G. Polymer (London) **15**, 366–372 (1974).

280b. Tonelli, A. E., Helfand, E.: Elastically ineffective crosslinks in rubbers. ASC Polymer Preprints **15**, 605–612 (1974).

281. Schultz, A. R.: Characterization of polymer networks, In: McIntyre, D. (Ed.): Characterization of macromolecular structure, pub. 1573, pp. 389–405. Washington: Nat. Acad. Sci. 1968.

282. Scanlan, J.: The effect of network flaws on the elastic properties of vulcanizates. J. Polymer Sci. **43**, 501–508 (1860).

283. Gordon, M., Kucharuk, S., Ward, T. C.: The statistics of elastically active network chains and the efficiency of crosslinking in rubbers. Collection Czech. Chem. Commun. **35**, 3252–3264 (1970).

283a. Tobolsky, A. V., Carlson, D. W., Indictor, N.: Rubber elasticity and chain configuration. J. Polymer Sci. **54**, 175–192 (1961).

284. Gordon, M., Ward, T. C., Whitney, R. S.: Chemical and physical aspects of the three stages in forming polymer networks. In: Polymer networks, structure, and mechanical properties, pp. 1–23. See Ref. (*260*).

285. Mancke, R. G., Dickie, R. A., Ferry, J. D.: Dynamic mechanical properties of cross-linked rubbers. V. An approximate analysis of the modulus contributions of trapped and untrapped entanglements. J. Polymer Sci., Part A-2 **6**, 1783–1789 (1968).

286. Kraus, G.: Quantitative characterization of polybutadiene networks. J. Appl. Polymer Sci. **7**, 1257–1263 (1963).

287. Pearson, D. S., Skutnik, B. J., Bohm, G. G. A.: Radiation crosslinking of elastomers. I. Polybutadienes. J. Polymer Sci. Polymer Phys. Ed. **12**, 925–939 (1974).

288. Bristow, G. M.: Relation between stress-strain behavior and equilibrium volume swelling for peroxide vulcanizates of natural rubber and cis-1,4-polyisoprene. J. Appl. Polymer Sci. **9**, 1571–1578 (1965).

289. Flory, P. J., Rabjohn, N., Shaffer, M. C.: Dependence of elastic properties of vulcanized rubber on the degree of crosslinking. J. Polymer Sci. **4**, 225–245 (1949).

290. Mullins, L.: Determination of degree of crosslinking in natural rubber vulcanizate. J. Appl. Polymer Sci. **2**, 1–7 (1959).

291. St. Pierre, L. E., Dewhurst, H. A., Bueche, A. M.: Swelling and elasticity of irradiated polydimethylsiloxanes. J. Polymer Sci. **36**, 105–111 (1959).

292. Langley, N. R., Polmanteer, K. E.: The role of chain entanglements in rubber elasticity. Am. Chem. Soc. Polymer Preprints **13**, 235–240 (1972).
293. Langley, N. R., Polmanteer, K. E.: The relation of elastic modulus to crosslink and entanglement concentrations in rubber networks. J. Polymer Sci. Polymer Phys. Ed. **12**, 1023–1034 (1974).
294. Langley, N. R., Ferry, J. D.: Dynamic mechanical properties of crosslinked rubbers. VI. Poly(dimethylsiloxane) networks. Macromolecules **1**, 353–358 (1968).
295. Kramer, O., Ty, V., Ferry, J. D.: Entanglement coupling in linear polymers demonstrated by networks crosslinked in states of strain. Proc. Natl. Acad. Sci. **69**, 2216–2218 (1972).
296. Kramer, O., Carpenter, R. L., Ty, V., Ferry, J. D.: Entanglement networks crosslinked in states of strain, paper presented at Society of Rheology Meeting, Montreal, October 1973. Part of this work is described in Entanglement networks of 1,2-polybutadiene crosslinked in states of strain. I. Cross-linking at 0° C. Macromolecules **7**, 79–84 (1974), by the same authors.
297. Baldwin, F. P., Ver Strate, G.: Polyolefin elastomers based on ethylene and propylene. Rubber Chem. Techn. **45**, 709–881 (1972).
298. Cohen, R. E., Tschoegl, N. W.: Dynamic mechanical properties of block copolymer blends—a study of the effects of terminal chains in elastomeric materials. I. Torsion pendulum measurements. Intern. J. Polymeric Mater. **2**, 49–69 (1972); II. Forced oscillation measurements. *Ibid* **2**, 205–223 (1973); III. A mechanical model for entanglement slippage. *Ibid* (in press).
299. Kramer, O., Greco, R., Ferry, J. D.: Viscoelastic properties of butyl rubber networks containing reptating polyisobutylene. Paper presented at the San Diego meeting of the American Physical Society, March, 1973.
300. Edwards, S. F.: Statistical mechanics with topological constraints. I. Proc. Phys. Soc. **91**, 513–519 (1967).
301. Schill, G.: Catenanes, rotaxanes and knots. New York: Academic Press 1971.
302. Prager, S., Frisch, H. L.: Statistical mechanics of a simple entanglement. J. Chem. Phys. **46**, 1475–1483 (1967).
303. Alexander-Katz, R., Edwards, S. F.: The statistical mechanics of entangled polymers. J. Phys. A Gen. Phys. **5**, 674–681 (1972).
304. Edwards, S. F., Freed, K. F.: The entropy of a confined polymer. I. J. Phys. A. Gen. Phys. **2**, 145–150 (1969).
305. Collins, R., Wragg, A.: The entropy of a confined polymer. II. J. Phys. A. Gen. Phys. **2**, 151–156 (1969).
306. Eichinger, B. E.: Elastic theory. II. Matrix method for a confined circular chain. J. Chem. Phys. **57**, 1356–1357 (1972).
307. Peterlin, A.: Rheology of polymer systems. J. Elastoplastics **4**, 112–130 (1972).
308. Ito, Y., Shishida, S.: Critical molecular weight for onset of non-newtonian flow and upper newtonian viscosity of polydimethylsiloxane. J. Polymer Sci., Polymer Phys. Ed. **10**, 2239–2278 (1972).
309. Philippoff, W.: Correlation of the elastic properties in steady-state flow and vibrational experiments. J. Appl. Phys. **36**, 3033–3038 (1965).
310. Bueche, F.: Influence of rate of shear on the apparent viscosity of *a*—dilute polymer solutions and *b*–bulk polymers. J. Chem. Phys. **22**, 1570–1576 (1954).
311. Pao, Y.-H.: Dependence of intrinsic viscosity of dilute solutions of macromolecules on velocity gradient. J. Chem. Phys. **25**, 1294–1295 (1956).
312. Onogi, S., Masuda, T., Kitagawa, K.: Rheological properties of anionic polystyrenes. I. Dynamic viscoelasticity of narrow-distribution polystyrenes. Macromolecules **3**, 109–116 (1970).
313. Penwell, R. C., Graessley, W. W., Kovacs, A.: J. Polymer Sci. Polymer Phys. Ed. (in press).
314. Cox, W. P., Merz, E. H.: Correlation of dynamic and steady flow viscosities. J. Polymer Sci. **28**, 619–622 (1958).
315. Stratton, R. A.: Non-newtonian flow in polymer systems with no entanglement coupling. Macromolecules **5**, 304–310 (1972).
316. Harris, E. K., Jr.: Viscometric properties of polymer solutions and blends as functions of concentration and molecular weight. Ph. D. thesis, University of Wisconsin, 1970.

317. Ito, Y., Shishido, S.: A modified Graessley theory for non-newtonian viscosity of poly-dimethylsiloxanes and their solutions. J. Polymer Sci. Polymer Phys. Ed. **12**, 617–628 (1974).

318. Noda, I., Yamada, Y., Nagasawa, M.: The rate of shear dependence of the intrinsic viscosity of monodisperse polymer. J. Phys. Chem. **72**, 2890–2898 (1968).

319. Kotaka, T., Suzuki, H., Inagaki, H.: Shear-rate dependence of the intrinsic viscosity of flexible linear macromolecules. J. Chem. Phys. **45**, 2770–2773 (1966).

320. Suzuki, H., Kotaka, T., Inagaki, H.: Shear-rate dependence of the intrinsic viscosity of flexible linear macromolecules. II. Solvent effect. J. Chem. Phys. **51**, 1279–1285 (1969).

321. Yamaguchi, N.: Sugiura, Y., Okamo, K., Wada, E.: Non-newtonian viscosity and excluded volume effect of dilute solutions of flexible high polymers. J. Phys. Chem. **75**, 1141–1149 (1971).

322. Gruver, J. T., Kraus, G.: Rheological properties of polybutadienes prepared by n-butyl-lithium initiation. J. Polymer Sci. Pt. A **2**, 797–810 (1964).

323. Lee, C. L., Polmanteer, K. E., King, E. G.: Flow behavior of narrow-distribution poly-dimethylsiloxane. J. Polymer Sci. Pt. A-2 **9**, 1909–1916 (1970).

324. Stratton, R. A.: The dependence of non-newtonian viscosity on molecular weight for monodisperse polystyrene. J. Colloid Sci. **22**, 517–530 (1966).

325. Paul, D. R., St. Laurence, J. E., Troell, J. H.: Flow behavior of concentrated solutions of an SBS block copolymer Polymer Eng. Sci. **10**, 70–78 (1970).

326. Mendelson, R. A., Bowles, W. A., Finger, F. L.: Effect of molecular structure on poly-ethylene melt rheology. I. Low shear behavior. J. Polymer Sci. Pt. A-2 **8**, 105–126 (1970).

327. Kataoka, T.: Onset of non-newtonian flow: a correlation with an average molecular weight. Polymer Letters **5**, 1063–1068 (1967).

328. Vinogradov, G. V.: Flow and rubber elasticity of polymer systems. Pure and Appl. Chem. **26**, 423–449 (1971).

329. Markovitz, H.: The reduction principle in linear viscoleasticity. J. Phys. Chem. **69**, 671 (1965).

330. Markovitz, H.: (Private communication).

331. Subirana, J. A.: Solvent effects on the non-newtonian viscosity of dilute solutions of flexible linear macromolecules. J. Chem. Phys. **41**, 3852–3856 (1964).

332. Peterlin, A.: Gradient dependence of intrinsic viscosity of freely flexible linear macromolecules. J. Chem. Phys. **33**, 1799–1802 (1960).

333. Burow, S. P., Peterlin, A., Turner, D. T.: The upturn effect in the non-newtonian viscosity of polymer solutions. Polymer (London) **6**, 35–47 (1965).

334. Fixman, M.: Polymer dynamics: non-newtonian intrinsic viscosity. J. Chem. Phys. **45**, 793–803 (1866).

335. Cottrell, F. R., Merrill, E. W., Smith, K. A.: Conformation of polyisobutylene in dilute solution subjected to a hydrodynamic flow field. J. Polymer Sci. Pt. A-2 **7**, 1415–1434 (1969).

336. Cottrell, F. R., Merrill, E. W., Smith, K. A.: Intrinsic viscosity and axial extension ratio of random-coiling macromolecules in a hydrodynamic flow field. J. Polymer Sci. Pt. A-2 **8**, 289–294 (1970).

337. Champion, J. V., Davis, I. D.: Light scattering by solutions of flexible macromolecules subjected to flow. J. Chem. Phys. **52**, 381–385 (1970).

338. Warner, H. R., Jr.: Kinetic theory and rheology of dilute suspensions of finitely extendible dumbbells. Ind. Eng. Chem. Fundam. **11**, 379–387 (1972).

339. Booij, H. C., van Wiechen, P. H.: Effect of internal viscosity on the deformation of a linear macromolecule in a sheared solution. J. Chem. Phys. **52**, 5056–5068 (1870).

339a. Zimmerman, R. D., Williams, M. C.: Evaluation of internal viscosity models. Trans. Soc. Rheol. **17**, 23–46 (1973).

339b. Bazúa, E. R., Williams, M. C.: Rheological properties of internal viscosity models with stress symmetry. J. Polymer Sci. Polymer Phys. Ed. **12**, 825–843 (1974).

340. Munk, P., Peterlin, A.: Streaming birefringence. IX. Invariant expressions for chain molecules. Rheol. Acta **9**, 288–293 (1970).

341. Ree, T., Eyring, H.: Theory of non-newtonian flow. I. Solid plastic system. J. Appl. Phys. **26**, 793–800 (1955).

342. Woods, M. E., Krieger, I. M.: Rheological studies on dispersions of uniform colloidal spheres. I. Aqueous dispersions in steady shear flow. J. Colloid Sci. **34**, 91–99 (1970).
343. Papir, Y. S., Krieger, I. M.: Rheological studies on dispersions of uniform colloidal spheres. II. Dispersions in non-aqueous media. J. Colloid Sci. **34**, 126–130 (1970).
344. Graessley, W. M.: Molecular entanglement theory of flow behavior in amorphous polymers. J. Chem. Phys. **43**, 2696–2703 (1965).
345. Graessley, W. W., Segal, L.: Effect of molecular weight distribution on the shear rate dependence of viscosity in polymer systems. A.I.Ch.E. J. **16**, 261–267 (1970).
346. Abdel-Alim, A. H., Balke, S. T., Hamielec, A. E.: Flow properties of polystyrene solutions under high shear rates. J. Appl. Polymer Sci. **17**, 1431–1442 (1973).
347. Bueche, F.: Viscosity of entangled polymers; theory of variation with shear rate. J. Chem. Phys. **48**, 4781–4784 (1968).
348. Shroff, R. N., Shida, M.: Correlation between steady state flow curve; and molecular weight distribution for polyethylene melts. Polymer Eng. Sci. **11**, 200–204 (1971).
349. Middleman, S.: Effect of molecular weight distribution on viscosity of polymeric fluids. J. Appl. Polymer. Sci. **11**, 417–424 (1967).
350. Philippoff, W., Gaskins, F. H., Brodnyan, J. G.: Flow birefringence and stress. V. Correlation of recoverable shear strains with other rheological properties of polymer solutions. J. Appl. Phys. **28**, 1118–1123 (1957).
351. Spriggs, T. W., Huppler, J. D., Bird, R. B.: An experimental appraisal of viscoelastic models. Trans. Soc. Rheol. **10**, 191–213 (1966).
352. Graessley, W. W., Segal, L.: Flow behavior of polystyrene systems on steady shearing flow. Macromolecules **2**, 47–57 (1969).
353. Tanaka, T., Yamamoto, M., Takano, Y.: Non-newtonian flow in concentrated polymer systems. J. Macromol. Sci. Pt. B **4**, 931–946 (1970).
354. Bird, R. B., Warner, H. R., Jr.: Hydrodynamic interaction effects in rigid dumbbell suspensions. I. Kinetic theory. Trans. Soc. Rheol. **15**, 741–750 (1971).
354a. Wales, J. L. S., Philippoff, W.: The anisotropy of simple shearing flow. Rheol. Acta **12**, 25–34 (1973).
355. Vinogradov, G. V., Belkin, I. M.: Elastic strength, and viscous properties of polymer (polyethylene and polystyrene) melts. J. Polymer Sci. Pt. A **3**, 917–932 (1965).
356. Huppler, J. D., MacDonald, I. F., Ashare, E., Spriggs, T. W., Bird, R. B., Holmes, L. A.: Rheological properties of three solutions. Part II. Relaxation and growth of shear and normal stresses. Trans. Soc. Rheol. **11**, 181–204 (1967).
357. Zapas, L. J., Phillips, J. C.: Simple shearing flows of polyisobutylene solutions. J. Res. Natl. Bur. St. **75A**, 33–40 (1971).
358. Schurz, J., Mavrommatakos, A.: Die prästationären Anlaufkurven bei Polyvinylacetat-Lösungen. Angew. Makromol. Chem. **15**, 95–107 (1971).
359. Schremp, F. W., Ferry, J. D., Evans, W. W.: Mechanical properties of substances of high molecular weight. IX. Non-Newtonian flow and stress relaxation in concentrated polyisobutylene and polystyrene solutions. J. Appl. Phys. **22**, 711–717 (1951).
360. Sakai, M., Fukaya, H., Nagasawa, M.: Time dependent viscoelastic properties of concentrated polymer solutions. Trans. Soc. Rheol. **16**, 635–649 (1972).
361. Einaga, Y., Osaki, K., Kurata, M., Kimura, S., Tamura, M.: Stress relaxation of polymers of polymers under large strain. Polymer J. (Japan) **2**, 550–552 (1971). See also Polymer J. (Japan) **5**, 91–96 (1973).
362. Berry, G. C.: (Private communication).
363. Osaki, K., Tamura, M., Kurata, M., Kotaka, T.: Complex modulus of concentrated polymer solutions in steady shear. J. Phys. Chem. **69**, 4183–4191 (1965).
364. Booij, H. C.: Influence of superimposed steady shear flow on the dynamic properties of non-Newtonian fluids. Rheol. Acta **5**, 215–221 (1966).
365. Simmons, J. M.: Dynamic modulus of polyisobutylene solutions in superposed steady shear flow. Rheol. Acta. **7**, 184–188 (1968).
366. Tanner, R. I., Williams, G.: On the orthogonal superposition of simple shearing and small-strain oscillatory motions. Rheol. Acta **10**, 528–538 (1971).
367. Sell, J. W., Forsman, W. C.: Loss moduli under steady shearing of concentrated polystyrene solutions. Macromolecules **5**, 23–24 (1972).

368. Lee,K.H., Jones,L.G., Pandalai,K., Brodkey,R.S.: Modifications of an R-16 Weissenberg Rheogoniometer.Trans. Soc. Rheol. **14**, 555–572 (1970).
369. Meissner,J.: Modification of the Weissenberg rheogoniometer for measurement of transient rheological properties of molten polyethylene under shear. Comparison with tensile data. J. Appl. Polymer Sci. **16**, 2877–2899 (1972).
370. Pipkin,A.C.: Small displacements superposed on viscometric flow. Trans. Soc. Rheology **12**, 397–408 (1968).
371. Markovitz,H.: Small deformations superimposed on steady viscometric flows. In: Onogi,S. (Ed.): Proc. 5th Internat. Cong. Rheology, Vol. I, pp. 499–510. Maryland: University Park Press 1970.
372. Bernstein,B., Fosdick,R.L.: On four rheological relations. Rheol. Acta **9**, 186–193 (1970).
373. Penwell,R.C.: (Private communication).
374. Lodge,A.S., Meissner,J.: Comparison of network theory predictions with stress/time data in shear and elongation for a low density polyethylene melt. Rheol. Acta **12**, 41–47 (1973).
375. Bagley,E.B., Storey,S.H., West,D.C.: Post extrusion swelling in polyethylene. J. Appl. Polymer Sci. **7**, 1661–1672 (1963).
376. Stratton,R.A., Butcher,A.F.: Stress relaxation upon cessation of steady flow and the overshoot effect of polymer solutions. J. Polymer Sci. Polymer Phys. Ed. **11**, 1747–1758 (1973).
377. vanEs,H.E., Christensen,R.M.: A critical test for a class of non-linear constitutive equations. Trans. Soc. Rheol. **17**, 325–328 (1973).
378. Bogue,D.C., White,J.L.: Engineering analyses of non-newtonian fluids. AGARDograph 144 (N.A.T.O.). Springfield, Va.: National Technical Information Service, 1970.
379. Carreau,P.J.: Rheological equations from molecular network theories. Trans. Soc. Rheol. **16**, 99–127 (1972).
380. Ziabicki,A.: Structural theories in polymer rheology. In: Proc. 5th Internat. Cong. Rheology, Vol. 3, pp. 235–253. Maryland: University Park Press 1970.
381. Tanner,R.I., Simmons,J.M.: Combined simple and sinusoidal shearing in elastic liquids. Chem. Eng. Sci. **22**, 1803–1815 (1967).
382. Ferry,J.D.: An example of a rheological conceptual scheme. In: Onogi,S. (Ed.): Proc. 5th Internat. Cong. of Rheology, Vol. 1, pp. 3–21. Maryland: University Park Press 1970.

Received May 28, 1974

Die

Makromolekulare

Chemie

An International Journal
of Macromolecular Chemistry and Physics

Founded by Hermann Staudinger, Nobel Prize Laureate in Chemistry
Editor: Werner Kern

Die Makromolekulare Chemie

- carries papers from authors working at all major universities and research centers throughout the world;

- publishes only original papers that meet high standards of excellence and significance;

- is clearly divided into: "Chemistry of Macromolecules", "Physical Chemistry of Macromolecules", and "Physics of Macromolecules";

- provides for speedier publication of important news as "Short Communications";

- gives "Summeries" in English as well as in the original language (65% of the papers being written in English, 25% in German and 10% in French).

Die Makromolekulare Chemie is published in 12 issues a year totalling about 3600 pages. Rate for 1974 (volume 175): sfr. 1260.– / DM 1056.–. Back issues and complete-back sets are available. Please inquire.

Hüthig & Wepf Verlag, Eisengasse 5, CH-4001 Basel

Advances in Polymer Science

Edited by H.J. Cantow,
G. Dall'Asta, J.D. Ferry,
H. Fujita, M. Gordon,
W. Kern, G. Natta,
S. Okamura,
C.G. Overberger, W. Prins,
G.V. Schulz, W.P. Slichter,
A.J. Staverman, J.K. Stille.

Advances in Polymer
Science comprises reports
of monograph type,
dealing with progress
achieved in the physics
and chemistry of high polymers, and includes full
bibliographical data. Its
objectives are to provide
those working in this field
with information on sub-
jects that are of special
topical interest and to
report recent advances that
have been so rapid as to
require review-type treatment.

Vol. 11:
56 figures. III, 204 pages
1973. Cloth DM 88,—;
US $35.90
ISBN 3-540-06054-5

Contents: R. Hayakawa,
Y. Wada: Piezoelectricity
and Related Properties of
Polymer Films. H. Tani:
Stereospecific Polymeri-
zation of Aldehydes and
Epoxides. H.-G. Elias,
R. Bareiss, J.G. Watterson:
Mittelwerte des Moleku-
largewichtes und anderer
Eigenschaften.

Vol. 12:
62 figures. III, 190 pages
1973. Cloth DM 78,—;
US $31.90
ISBN 3-540-06431-1

Contents: K. Osaki:
Viscoelastic Properties of
Dilute Polymer Solutions.
W.L. Carrick: The Mecha-
nism of Olefin Polymeri-
zation by Ziegler-Natta
Catalysts. C. Tosi,
F. Ciampelli: Applications
of Infrared Spectroscopy
to Ethylene-Propylene
Copolymers. K. Tsuji:
ESR Study of Photode-
gradation of Polymers.

Vol. 13: W. Wrasidlo
**Thermal Analysis of
Polymers**
49 figures. II, 99 pages
1974. Cloth DM 46,—;
US $18.80
ISBN 3-540-06552-0

Contents: Scope of the
Review. Experimental
Methods. Glass Transi-
tions. Melting. Crystallization.

Vol. 14:
11 figures. III, 130 pages
1974. Cloth DM 58,—;
US $23.70
ISBN 3-540-06649-7

Contents: J.P. Kennedy,
S. Rengachary: Correlation
between Cationic Model
and Polymerization Reac-
tions of Olefins.
J. Hutchison, A. Ledwith:
Photoinitiation of Vinyl
Polymerization by Aroma-
tic Carbonyl Compounds.

L.S. Gal'braikh,
Z.A. Rogovin: Chemical
Transformations of
Cellulose.

Vol. 15
32 figures
Approx. 140 pages. 1974
Cloth DM 68,—; US $27.80
ISBN 3-540-06910-0

Contents: G. Henrici-Olivé,
S. Olivé: Oligomerization
of Ethylene with Soluble
Transition-Metal Catalysts.
A. Zambelli, C. Tosi:
Stereochemistry of Propy-
lene Polymerization.
C.-D.S. Lee, W.H. Daly:
Mercaptan-Containing
Polymers. Yu. V. Kissin:
Structures of Copolymers
of High Olefins.

Prices are subject to
change without notice

**Springer-Verlag
Berlin
Heidelberg
New York**
München Johannesburg
London Madrid
New Delhi Paris
Rio de Janeiro Sydney
Tokyo Utrecht Wien